Study Guide

for

Andersen and Taylor's

Sociology
The Essentials

Study Guide

for

Andersen and Taylor's

Sociology
The Essentials

Fourth Edition

Jan Abu-Shakrak
Portland Community College

THOMSON
™
WADSWORTH

Australia • Brazil • Canada • Mexico
Singapore • Spain • United Kingdom • United States

Printer: Thomson West

ISBN 0-495-09297-5
Cover image: © Vivienne Flesher

Thomson Higher Education
10 Davis Drive
Belmont, CA 94002-3098
USA

For more information about our products, contact us at:
Thomson Learning Academic Resource Center
1-800-423-0563

For permission to use material from this text or product, submit a request online at
http://www.thomsonrights.com.
Any additional questions about permissions can be submitted by email to **thomsonrights@thomson.com.**

TABLE OF CONTENTS

PREFACE

This study guide accompanies the fourth edition of *Sociology: The Essentials* by Margaret Andersen and Howard Taylor. Used as a supplement to the textbook, this guide can assist you in learning course material by highlighting core concepts, key people, principal theories, and relevant research on a variety of interesting topics in sociology.

Each chapter in the study guide begins with a brief chapter outline, a chapter focus statement, and a set of questions to guide your reading. These questions direct your attention to some of the most important issues covered in the corresponding chapter of the textbook.

For each chapter, you should be able to: define and give an example of the key terms presented; identify the main contributions of the key people whose work is discussed in the chapter; and apply each theoretical perspective to the topics discussed in the textbook. The outline for each chapter highlights the issues discussed in the text, and can be used either as a review guide or as a summary framework to focus your reading.

To examine specific issues further, do the Internet and InfoTrac Exercises. Your course instructor may assign selected exercises.

You can evaluate your comprehension of the material by completing the practice tests located at the end of each chapter in the study guide. Each practice test includes multiple choice, true-false, fill in the blank, and essay questions, followed by page number references, a detailed answer key, and a guide to help you write responses to the essay questions.

I hope this Study Guide enhances your reading of the textbook and your learning process.

Jan Abu-Shakrah
Portland Community College
Portland, Oregon

January 2006

SOCIOLOGICAL PERSPECTIVES AND SOCIOLOGICAL RESEARCH

BRIEF CHAPTER OUTLINE

Evaluation Research

Research Ethics: Is Sociology Value-Free?

Chapter Summary

CHAPTER FOCUS

This chapter provides an overview of the discipline of sociology and its historical development, sociological theory, and sociological research.

QUESTIONS TO GUIDE YOUR READING

1. What is sociology, and how does it differ from the other social sciences?

2. What is the sociological perspective, and what is the role of debunking in understanding human behavior and social change?

3. How are social structure, social institutions, social change, and social interaction defined, and what role do they play in grasping the sociological perspective?

4. What do sociologists mean by diversity?

5. In what context did sociology develop, and what were the contributions of the giants of European sociology—Durkheim, Marx, and Weber?

6. What are the major contributions and characteristics of American sociology, and who were the early American sociologists?

7. What are the three major theoretical frameworks within sociology, and what core assumptions do theorists associated with each framework make about society?

8. What are the stages of the sociological research process?

9. What are the major tools or methods sociologists use to gather data, and what are the advantages and limits of each tool?

10. Can sociology be value-free?

KEY TERMS

(defined at page number shown and in glossary)

conflict theory 17	**debunking** 6
controlled experiment 25	**dependent variable** 22
data 22	**empirical** 5

KEY PEOPLE

(identified at page number shown)

CHAPTER OUTLINE

I. WHAT IS SOCIOLOGY?

A. **Sociology** is the scientific study of human behavior and the social context in which it occurs. Academic disciplines concerned with social behavior and social change make up the *social sciences*. For sociologists, the unit of analysis is groups or whole societies, whereas *psychology* analyzes individual behavior. *Anthropology* is the study of human culture as the basis for society. *Political science* is the study of politics and the organization of government. *Economics* is the study of the production, distribution, and consumption of goods and services. Sociologists view culture and specific social institutions as interrelated parts of a complex set of systems that comprise society. *Social work* is an applied field that draws on the lessons of all the social sciences to serve people in need.

A. The Sociological Perspective

    1. The *sociological perspective* allows you to see the societal patterns that influence individual and group life.

1. C. Wright Mills (1916-1962) explored the concept of the **sociological imagination**, the ability to identify the societal patterns that influence individual and group life. He emphasized the importance of understanding the social and historical context in which people live. Mills distinguished between **troubles** – privately felt problems that develop from events or emotions in an individual's life, and **issues** – problems that affect large numbers of people and are based in the history and institutional arrangements of society. This distinction is the crux of the difference between individual experience and **social structure**.

2. Sociology is an **empirical** discipline where rigorous methods of research are used to investigate everyday life, and conclusions must be based on careful, systematic observations, rather than personal opinion.

B. Discovering Inconvenient Facts

    1. Sociological research provides evidence that persistent problems, such as women's lower earnings relative to men's, are embedded in society, not just individual behavior. Sociology often reveals troubling facts about society; for example, the United States has the highest rate of imprisonment in the world. It also asks intriguing questions about everyday life, and looks at the odd corners of society.

C. Debunking in Sociology

1. Peter Berger used the term **debunking** to refer to sociology's "unmasking tendency," or its role in looking beyond the facades of everyday life. For example, sociology helps reveal situations in schools whereby the opportunities of some children are stifled, in contrast to the common belief that all children are given the opportunity to learn and succeed.

2. Debunking is sometimes easier to do when examining cultures other than one's own because one's thinking is less restricted by the common assumptions within the culture. For example, Westerners may view the traditional Chinese custom of footbinding as bizarre, but the practice of enhancing women's breasts by implanting silicon may seem normal.

3. Elaine Bell Kaplan's investigation of the social structural context of Black teen mothers' lives debunks common stereotypes and myths about this group.

D. Key Sociological Concepts

1. **Social structure** is the organized pattern of social relationships and social institutions that together constitute society. Social forces guide human behavior.

2. **Social institutions** are established, organized systems of social behavior with particular, recognized purposes. Examples of major institutions are the family, religion, government, and the economy.

3. Sociologists do not view society as fixed or immutable; thus, they are interested in the process of **social change,** or the alteration of society over time.

4. **Social interaction**, which is behavior between two or more people, is given meaning by the participants, illustrating that people are active agents in society.

II. THE SIGNIFICANCE OF DIVERSITY

A. Defining Diversity

1. **Diversity** is a broad concept that refers to the variety of group experiences that result from the social structure of society. Although race, class, and gender are critical components of diversity in the United States, factors such as age, nationality, sexual orientation, and region also influence people's identities, experiences, and opportunities.

B. Society in Global Perspective

1. Nations are increasingly connected to each other in complex ways, making it necessary to use a global perspective to understand topics such as employment and immigration. For example, a Latina woman may leave her child in her homeland to go to the United States to work as a domestic for an American woman who is pursuing her career.

III. THE DEVELOPMENT OF SOCIOLOGICAL THEORY

A. Sociology first emerged in the eighteenth and nineteenth centuries, when rapid political and economic changes were occurring in Western Europe.

A. The Influence of the Enlightenment

1. The **Enlightenment**, known as the Age of Reason, strongly influenced the development of sociology. The Enlightenment's faith in the ability of human reason to solve society's problems by identifying natural social laws and processes was strongly linked to the development of modern science.

2. **Auguste Comte** (1798-1857), who first coined the term *sociology*, believed that the discipline was the most evolved of all sciences because it studied the complex and changing systems of

society. His approach is called **positivism**, a system of thought in which accurate observation and description is considered the highest form of knowledge.

    3. **Alexis de Tocqueville** (1805-1859), author of *Democracy in America*, explored democratic and egalitarian values in the United States, which he argued influenced social institutions and personal relationships. He believed that individualism made people self-centered and anxious about their position.

    4. **Harriet Martineau** (1802-1876) was a British feminist and abolitionist who analyzed social customs in the United States in her book, *Society in America*. She wrote the first sociological research methods book, *How to Observe Manners and Morals*, which examined the technique of participant observation.

B. <u>Classical Sociological Theory</u>

    1. **Emile Durkheim** (1858-1917) explored *social solidarity*, or the bonds that link members of a group. For example, he believed that religious rituals reinforce the sense of belonging of group members, and members express social solidarity by condemning deviant behavior.

       a. Durkheim discussed society *sui generis*, referring to his belief that society is a subject that should be studied separately from the total of individuals who comprise it.

       b. Durkheim's work was the basis for *functionalism*, one of the major theories of sociology. He viewed society as an integrated whole, with each part contributing to the stability of the system.

       c. Durkheim noted that **social facts**, the social patterns that are *external* to individuals (such as values), were the proper subject of sociology.

    2. **Karl Marx** (1818-1883) investigated the effects of capitalism on individuals and societies, arguing that all other social institutions (family, law, education) are shaped by economic forces.

       a. Capitalism is an economic system based on profit and private property, where the capitalist class owns the means of production.

       b. Capitalists gain profit through the exploitation of the proletariat, or working class, who sell their labor in exchange for wages.

       c. Marx's work is often misinterpreted and many of his predictions did not occur, but he produced an important body of work that identified class as a fundamental dimension of society.

    3. **Max Weber** (1864-1920) developed a *multidimensional* analysis of society that recognized the interplay between three basic institutions -- political, economic, and cultural.

       a. Weber theorized that there could not be a value-free sociology. Rather, he thought sociologists should acknowledge the influence of their values and beliefs so they would not interfere with their objectivity.

       b. Weber advocated that sociologists use the concept of **verstehen** in their work. This would facilitate a more subjective understanding of social behavior from the point of view of those engaged in it.

       c. Weber defined *social action* as behavior to which people give meaning. It is the responsibility of sociologists to identify those meanings and the context in which behavior occurs.

C. <u>Sociology in America</u>

    1. Although sociology in the United States developed from the foundations of the discipline begun in Europe, American sociology had some unique characteristics. *Pragmatism*, a belief in practicality, led sociologists to value social planning. There was an emphasis on identifying the causes of social problems and developing strategies to improve them.

    1. Early sociologists viewed society as an organism, a system of interrelated parts that work together. This is referred to as the **organic metaphor.**

    2. **Social Darwinism** adapted **Charles Darwin**'s (1809-1882) theory of biological *evolution* to analyze social change, asserting that societies evolve along a natural course; thus, this group supported a *laissez-faire* (hands-off) approach to society. Most other early U.S. sociologists took a more reform-based approach.

    3. Two key aspects of the Chicago School of sociology were an interest in how society shaped the mind and identity of individuals, and the use of social settings as human laboratories for research.

        a. **Charles Horton Cooley** (1864-1929) believed that an individual's identity is based on his/her understanding of how others perceive them.

        b. **George Herbert Mead** (1863-1931) extended Cooley's idea by investigating how individuals develop through the relationships they establish with others.

        c. Sociologists **W. I. Thomas** (1863-1947) and **Florian Znaniecki** (1882-1958) used Polish immigrants' personal documents to examine their social relationships and feelings about their new lives in the U.S.

        d. **Robert Park** (1864-1944) investigated interaction between people of different races as well as the sociological design of cities. He developed the concentric circle model of urbanization.

        e. Although sexism prevented most female sociologists from securing academic positions, many women, including **Jane Addams** (1860-1935), entered the applied field of social work. Addams won the Nobel Peace Prize in 1931 for her work in the settlement house movement.

    4. African American sociologists have a long tradition of investigating Black communities, despite their historical exclusion from White universities. **W. E. B. Du Bois** (1868-1963) was the first Black person to receive a doctorate from Harvard in any field. Du Bois viewed sociology as a scientific, community-based, activist profession committed to social justice. Du Bois co-wrote *The Philadelphia Negro*, one of the first empirical community studies published.

D. <u>Theoretical Frameworks in Sociology: Functionalism, Conflict Theory, and Symbolic Interaction</u>

    1. Sociologists use theories to organize observations, produce logically related statements about observed behavior, and relate their observations to other data. Sociological theories use one of two basic approaches. In the m*acrosociology* realm are theories, such as those developed by Durkheim, Marx, and Weber, which seek to understand society as a whole. *Microsociology*, including theories, including those developed in the Chicago School, focus on face-to-face interaction. The three major sociological perspectives are functionalism, conflict theory, and symbolic interaction [Table 1.1].

1. **Functionalism** investigates how each part of society contributes to the stability of the whole system. Functionalist theorists emphasize order in society, noting that *disorganization* in the system leads to change. Because all parts of a society are related, a change in one part leads to changes throughout the society.
   a. **Talcott Parsons** (1902-1979) identified the four principal functions of society as: adaptation to the environment, goal attainment, integration of members into harmonious units, and maintenance of basic cultural patterns.
   b. **Robert Merton** (1910-2003) realized that social practices can have unintended consequences that are neither immediately apparent nor necessarily the same as their stated purposes, called **latent functions**. He distinguished these from **manifest functions**, or the stated, intended goals of social behavior.
   c. Critics of functionalism argue that this theory offers a conservative view of society and understates power differences.
2. Conflict theory emphasizes the role of coercion in producing social order, noting that **power** is the ability to influence and control others.
   a. This theory, based on Karl Marx's view of society as fragmented into groups that compete for social and economic resources, asserts that groups with the most resources utilize their power to defend and maintain their advantages in society.
   b. Conflict theorists view inequality as inherently unfair and believe social change is the result of struggles between competing groups.
   c. Critics argue that this theory neglects shared values and understates the degree of cohesion and stability in society.
3. **Symbolic interaction theory** views social interaction as the basis of society. It is used to investigate face-to-face interactions as a way of identifying the subjective meanings that people give to objects, events, and behavior.
   a. People behave based on what they *believe,* not just what is objectively true; thus, society is highly subjective.
   b. Society is "socially constructed" through human interpretation and constantly modified through social interaction.
   c. Critics of social interaction theory argue that it has a weak analysis of inequality and understates the objective basis of society.

E. Diverse Theoretical Perspectives

1. Sociological thought is diverse and includes other frameworks, like **exchange theory** and **rational choice theory**. **Postmodernism** argues that society is reflected in the words and images, or *discourses*, that people use to represent behavior and ideas.

IV. **DOING SOCIOLOGICAL RESEARCH**

The research methods sociologists use depend upon the questions they ask. For *Sidewalk*, for example, Mitch Dunier wanted to know how homeless people lived and so used **participant observation**. Other kinds of research on homelessness might ask different questions requiring use of official records or survey methods.

I. Sociology and the Scientific Method

A. Sociological research is derived from the **scientific method** and observes behavior and tests theory. Sociological insights can emerge from either **deductive** or **inductive** reasoning.

A. The Research Process
1. **Developing a Research Question.** The first step in sociological research is to develop a research question, which typically involves reviewing the existing literature on the subject. Some researchers conduct **replication studies**, which repeat a prior investigation exactly, but on a different group of people in a different time or place.
2. **Creating a Research Design.** A **research design** is the overall logic and strategy underlying a study. The design is created after the research question is developed because the technique to be used and other details of the design flow from the questions of the study.
   a. **Quantitative research** uses statistical procedures to analyze numerical data.
   b. **Qualitative research**, which is somewhat less structured and more interpretive than quantitative research, does not make extensive use of statistical methods, but tends to have greater depth.
   c. Some research designs involve the testing of a **hypothesis**, which is a prediction or tentative assumption that one plans to investigate. Hypotheses are formulated as "if-then" statements.
   d. *Exploratory research*, which is more open-ended, does not follow this model of hypothesis testing.
   e. Any research design includes a plan for how **data** will be gathered.
   f. Sociological research often investigates the influence of one **variable** on another. Variables are conditions or characteristics that can have more than one value, such as income.
      i. The **independent variable** is the one that the researcher wants to test as the presumed cause of something else.
      ii. The **dependent variable** is the one upon which there is a presumed effect.
      iii. Variables used to show more abstract concepts (like social class) are called **indicators**.
   g. The **validity** of a measurement is the degree to which it accurately measures or reflects a concept. To ensure validity, researchers usually measure more than one indicator for a particular concept.
   h. A measure is **reliable** if a repeat of the measurement gives the same result. Researchers often rely on measures that have proven sound in past studies to ensure reliability.
3. **Gathering Data.** After the research design is created, data are collected. Sociologists may gather their own original material, known as *primary data*; or use information that was previously collected and organized by another party (such as the Bureau of the Census), known as *secondary data*. A **population** is a relatively large collection of people or other units that a researcher studies and about which generalizations can be made. When it is impossible to include every member of the population in the study, the researcher selects a **sample**, or subset of the population, for the study. Researchers attempt to select samples that are *representative* and large enough to overcome statistical abnormalities to avoid selecting a *biased* sample. The best way to ensure a representative sample is to use a **random sample**, which gives everyone in the population an equal chance of being selected.

4. **Analyzing the Data. Data analysis** is the process by which sociologists organize and examine the data collected to search for patterns. Depending on the question asked and the type of data collected, statistical procedures may or may not be used to analyze the data.

5. **Reaching Conclusions and Reporting Results.** The final stage in the research process involves reaching conclusions and reporting the results. Sociologists attempt to draw conclusions from specific data and apply them to a broader population, a process called **generalization**.

II. **THE TOOLS OF SOCIOLOGICAL RESEARCH**

A. The Survey: Polls, Questionnaires, and Interviews

1. Surveys, which may be conducted by mail, in person, or over the telephone, are among the most commonly used tools of sociological research.

   a. In a closed-ended questionnaire, people choose a reply from the list of possible answers, similar to a multiple choice test.

   b. In an open-ended survey, respondents can provide their own responses, similar to an essay test.

2. Questionnaires are typically distributed to large numbers of people. The *return rate* refers to the percentage of questionnaires returned out of the total number distributed. A low return rate introduces possible bias in the results.

3. Surveys can include questions about a wide variety of topics and results can be efficiently analyzed to identify relationships among many different variables. However, respondents may disguise their true opinions if their answers are not considered socially acceptable.

B. Participant Observation

1. When using participant observation, the research becomes a member of the group they are studying, acting as both subjective participant and objective observer. This method is also known as *field research.*

2. Judith Rollins' study of maids and their employers is an example of a participant observation study that reveals how social class, race, and gender shape the experiences of Black domestic workers.

3. Participant observation is time-consuming and involves extensive note taking that generates a massive quantity of data that must be organized and examined. Because it involves investigating small groups, it may be difficult for the researcher to generalize the results of a participant observation study.

C. Controlled Experiments

1. **Controlled experiments** are highly focused ways of collecting data that are especially useful for determining a pattern of cause and effect.

2. To conduct a controlled experiment, two groups with similar subjects are created: the *experimental group*, which is exposed to the factor being investigated, and the control group, which is not exposed to the factor in question.

3. Although experiments can clearly establish causation, they often occur in artificial environments that eliminate some elements of real life, making it difficult to determine how much the laboratory setting affected the results.

D. Content Analysis

1. **Content analysis**, a method of investigating society and social behavior by examining cultural artifacts (including magazines, television commercials, novels, movies, and songs), is often used to indirectly determine how social groups are perceived over time.

2. This method has the advantage of being unobtrusive, yet it is limited to topics that are found in forms of mass communication. It only reveals how groups are depicted, not what people think about the images or how the images affect them.

E. Historical Research

    1. Historical research, a type of qualitative research, is commonly used to investigate people's life experiences. This type of research relies on data contained in historical archives, such as government and church records as well as private diaries and letters.

F. Evaluation Research

    1. **Evaluation research** assesses the effects of policies and programs on people in society. When the research is intended to produce policy recommendations, it is called *policy research*. *Market research* is another form of evaluation research where the sales potential of a product or service is determined by assessing customers' preferences.

III. **RESEARCH ETHICS: IS SOCIOLOGY VALUE-FREE?**

A. Sociologists, who often study controversial topics, do not usually claim to be value-free, but they try to produce objective research. Doing research raises ethical questions, such as the issue of deception in controlled experiments, where the actual purpose of the study is usually concealed. In participant observation studies, where the researcher may conceal his or her identity and not reveal that s/he is doing research, informed consent is not obtained from the people being studied. The professional code of ethics is clear that if a research subject is at risk of physical, mental, or legal harm, the subject must be informed of his or her rights and the researcher's responsibilities.

INTERNET EXERCISES

1. Visit the website of the American Sociological Association (www.asanet.org/), and investigate the ways in which students can get involved in the ASA through the student forum, the honors program, student travel, and other resources. Read through the online booklet, Careers in Sociology. Write a brief note explaining what you learned about Careers in Sociology. If you found a career that interests you, explain briefly what steps could advance the possibility of getting started in such a career.

2. Visit the website of the Society for the Study of Social Problems (www.sssp1.org). Look at the description of the society and its committees. Examine the links to other sociological sites of interest, as well as the most recent edition of the SSSP Newsletter. Write a brief note on what you found most interesting about this site or a resource on the site to share with other students in class.

INFOTRAC EXERCISES

Conduct a keyword search using InfoTrac College Edition to extend the discussion in DOING SOCIOLOGICAL RESEARCH: *Debunking the Myths of Black Teenage Motherhood* (p. 6-7)

1. **Keywords: Poverty and adolescence.** A search of refereed sources using these keywords will access articles linking poverty and adolescence to teen pregnancy and other risk factors such as dropping out of school.

2. **Keyword: Black motherhood.** This search will lead you to reviews of books on the topic, as well as research on risk factors for the children, male exploitation, and mothers in prisons.

3. **Keyword: Teen mothers.** This search leads to articles on welfare reform and work, risk factors for the children, and social control of adolescent sexuality.

4. **Keyword: Teen pregnancy.** This search leads to articles on teen pregnancy and parenting, the effectiveness of sex education, safe sex programs, the effectiveness of teen pregnancy programs, and the replacement of the "welfare queen" with the "exploited teen" myth to justify program assumptions about statutory rape and welfare dependency.

PRACTICE TEST

Multiple Choice Questions

1. Which of the following is **not** a main focus of sociology as a discipline?
   a. the ways that social institutions shape people's lives
   b. the contributions of famous individuals to society and history
   c. how attitudes and behaviors are socially patterned
   d. the influence of social change on people's experiences

2. Mills referred to the ability to identify the societal patterns that influence individual and group life as the _____.
   a. sociological imagination
   b. sociological structure
   c. debunking dynamic
   d. naturalizing attitude

3. Which of the following is **not** one of the primary professional activities engaged in by most sociologists?
   a. teaching in colleges and universities
   b. conducting research and working on public policy issues
   c. consulting with community groups working toward change
   d. providing therapeutic counseling to individuals under stress

4. Sociology is an empirical discipline, which means that _____.
   a. conclusions about behavior must be based on careful, systematic observations
   b. common sense is a reliable source of information for understanding behavior
   c. research on human behavior must occur in its natural setting
   d. all of these choices

5. Elaine Kaplan Bell's study of teenage pregnancy in Black communities indicates that _____.
   a. most adolescent girls intentionally become pregnant and have a baby to become eligible for welfare benefits that make them economically self-sufficient
   b. most adolescent girls who have a baby do not feel any stigma or shame about receiving welfare benefits because they believe the government owes them some assistance
   c. most mothers of adolescent girls who have babies are disappointed by their daughters' pregnancies because they think it will limit their opportunities
   d. most mothers of adolescent girls who have babies condone their daughters' early sexual behavior because they are happy that their daughters have boyfriends

6. Shirley Hill's study revealed that many African-American women became mothers even though doctors told them that they carried the gene that can cause sickle-cell anemia. Many of the women believed the medical warnings were unreliable and had children anyway because personal beliefs, rather than objective facts, motivate human behavior. This case illustrates the principles of _____ theory.
   a. conflict
   b. functionalist
   c. social exchange
   d. symbolic interaction

7. Which of the following "unsettling facts" about the United States is **true**?
   a. The rate of imprisonment is lower than in all other nations.
   b. Women and men with college degrees now earn the same average incomes.
   c. The poverty rate among White citizens is higher than that of Asian immigrants.
   d. Infant mortality, especially among African Americans and Hispanics, exceeds that of many impoverished nations.

8. The idea that sociology can be used to intervene in the natural evolution of society for the improvement of society is social _____.
   a. posivitism
   b. darwinism
   c. organicism
   d. telesis

9. Emile Durkheim's research on social solidarity illustrates the principles of which social theory?
   a. conflict
   b. functionalist
   c. social exchange
   d. symbolic interactionist

10. Max Weber stated that sociologists should try to understand human behavior from the perspective of the people engaged in it. He referred to this concept as _____.
    a. wohnen
    b. verstehen
    c. social solidarity
    d. debunking

11. The concept of debunking refers to _____.
    a. questioning the taken-for-granted assumptions of social life
    b. treating the people being studied as objects to avoid being biased
    c. studying only people who are similar to you to avoid misunderstandings
    d. making moral judgments about the beliefs and practices of the people being studied

12. Social Darwinism suggested that _____.
    a. sociologists should use research to improve the quality of life for all citizens
    b. social inequality is the result of the unequal distribution of economic resources
    c. sociology is not a real science because it does not use objective research methods
    d. society follows a natural evolutionary course in adapting to the environment

13. Joe was fired from his job because he repeatedly overslept and came to work late. This example illustrates the type of problem C. Wright Mills referred to as _____.
    a. deficiencies
    b. troubles
    c. issues
    d. faults

14. Two hundred men were laid off from their auto assembly jobs because the company closed the factory to cut costs. This example illustrates the type of problem C. Wright Mills referred to as _____.
    a. deficiencies
    b. troubles
    c. issues
    d. faults

15. Sociologists affiliated with the Chicago School of American sociology were _____.
    a. involved in using social settings as laboratories for human research
    b. interested in how society shaped the mind and identities of individuals
    c. committed to the application of sociological ideas to real social problems
    d. all of these choices

16. Which of the following statements about Jane Addams is **true**?
    a. She was an early sociologist of the Chicago School, but never held a regular teaching job.
    b. She was a leader in the settlement house movement.
    c. She was the only practicing sociologist ever to win a Nobel Peace Prize.
    d. all of these choices

17. Robert Merton realized that some social practices have consequences that are neither apparent nor consistent with their stated purpose. These unintended consequences of behavior are called _____.
    a. manifest functions
    b. latent functions
    c. social issues
    d. social facts

18. Dr. Smith is conducting a participant observation study in a college dormitory. She observes how the residents interact with each other and asks them how they feel about their living situation. The type of data she is collecting is _____.
    a. quantitative
    b. qualitative
    c. secondary
    d. predictive

19. The theoretical perspective that suggests society is reflected in the words and images that people use to represent behavior and ideas is called _____.
    a. feminism
    b. functionalism
    c. neodarwinism
    d. postmodernism

20. A teacher thinks that gender may cause differences in her students' performance on math tests. In this example, gender would be what kind of variable?
    a. hypothetical
    b. independent
    c. secondary
    d. dependent

21. Which of the following aspects of Judith Rollins' research on Black domestic workers raised potential ethical concerns?
    a. Rollins took a job as a maid and organized the other domestic workers to demand higher wages, eventually leading a strike against their White employers.
    b. Rollins took the role of an employer, hired several Black women as maids, and paid them without reporting their wages to the Internal Revenue Service.
    c. Rollins took the role of an employer and interviewed Black women for the job of maid but never intended to hire anyone.
    d. Rollins took a job as a maid without telling her employer that she was conducting research.

22. The disadvantage(s) of using surveys to investigate people's opinions on social issues is (are) _____.
    a. this method of collecting and analyzing data is especially time-consuming
    b. researchers can only study the views of a small number of people at one time
    c. respondents may use deception to conceal opinions that they believe are unacceptable
    d. all of these choices

23. Dr. Lee wants to know how the portrayal of women in magazines has changed over the last two decades. The most useful technique for her to use in this investigation is _____.
    a. controlled experiment
    b. program evaluation
    c. content analysis
    d. field research

24. The statistical tool used by researchers that is not skewed or distorted by extreme scores at either end of the distribution is the _____.
    a. mean
    b. mode
    c. median
    d. percentage

25. The statement, "If a man drops out of high school, then he is more likely to be unemployed than a man who graduates from high school," is an example of a _____.
    a. sample
    b. concept
    c. variable
    d. hypothesis

True-False Questions

1. Karl Marx considered all of society—laws, family structures, schools, and other institutions—to be shaped by economic forces.

   TRUE or FALSE

2. In *How to Observe Manners and Morals*, sociologist Harriet Martineau argued activists' efforts to grant women the vote in the United States threatened the stability of the American family.

   TRUE or FALSE

3. W. I. Thomas' famous saying, "If men define situations as real, they are real in their consequences," is referred to as the "irrational condition."

   TRUE or FALSE

4. Social Darwinists supported social telesis, the idea that society was best left alone to follow its natural evolutionary course toward perfection.

   TRUE or FALSE

5. There are larger concentrations of African American people in California and Texas than in any place in the United States.

   TRUE or FALSE

6. Auguste Comte, who first coined the term *sociology*, believed that sociology was the most highly evolved of all the sciences because it involved the study of the entire society.

    TRUE or FALSE

7. Robert Merton developed the idea that social institutions may serve two kinds of functions—latent and manifest.

    TRUE or FALSE

8. The overall logic and strategy underlying a research project is called the hypothetical map.

    TRUE or FALSE

9. Numerical data, such as those collected by the United States Bureau of the Census, are useful for doing quantitative research.

    TRUE or FALSE

10. Sociologists are ethically and legally prohibited from using deception in research and must always tell the subjects being studied the purpose of the research

    TRUE or FALSE

Fill-in-the-Blank Questions

1. Emile Durkheim conceptualized social _____ as those values, customs, and other patterns of behavior that are external to individuals and therefore, the proper subject of sociology.

2. Dr. Brown wants to know if the particular measure she is using accurately reflects the concept she is studying; thus, she is interested in the _____ of the measure.

3. Dr. Wood selected his sample in such a way that everyone in the population he was studying had an equal chance of being included, resulting in a _____ sample.

4. When a sociologist repeats a prior investigation with a different group of subjects in a different time or place, it is referred to as a(n) _____ study.

5. A researcher has a hunch that family income influences a person's likelihood of going to college. In this example, attending college would be a _____ variable.

Essay Questions

1. Distinguish between *troubles* and *issues* using C. Wright Mills' definitions of these concepts. Give an example of a social problem that appears to have an individual cause, but can be explained in a different way using the sociological imagination.

2. Define *diversity* and identify the three most influential aspects of diversity in the United States today. Explain why sociologists think examining diversity is crucial for understanding society.

3. Describe and give an example of four common statistical mistakes that researchers make when analyzing or interpreting data.

4. You are a sociologist who wants to investigate the topic of alcohol consumption on college campuses. Describe what steps you would take to develop the research question and design for this study. Discuss the advantages and disadvantages of using a survey, participant observation, and a controlled experiment to conduct this study.

5. You are a sociologist who is interested in studying school violence. Describe how you might approach the study of this phenomenon as a functionalist, a conflict theorist, and a symbolic interactionist.

Solutions

PRACTICE TEST

Multiple Choice Questions

1. B, 9-10 Sociologists study social institutions, socially patterned attitudes and behaviors, and patterns of social interaction, but the focus of sociology is not on the unique contributions of famous people.

2. A, 4-5 The sociological imagination is a way of looking at society that reveals social patterns. Social structure is the organized pattern of social relationships and social institutions that constitute society

3. D, 8 A majority of sociologists work in academic settings as teachers and researchers, and some work as consultants, but few provide individual counseling.

4. A, 5 Sociology is an empirical science because it is based on careful, systematic observations. Common sense is an unreliable source of information. Sociological research can occur in natural or controlled environments, but must be based on careful and systematic observation.

5. C, 6-7 Kaplan's research indicates that the common assumptions made about Black teenage mothers are myths. Her study showed that most of the mothers are disappointed and most of the girls feel shame about being pregnant and receiving welfare assistance.

6. D, 20 Hill's study illustrates the principles of symbolic interaction theory, which focuses on how people interpret information and attribute meanings to situations.

7. D, 6 Infant mortality, especially among African Americans and Hispanics, exceeds that of many impoverished nations. Women's average earnings are less than men's at all education levels. Asian immigrants have higher rates of poverty than do Whites. The rate of imprisonment in the United States is higher than in all other nations in the world.

8. D, 16 Lester Frank Ward supported social telesis, the idea that sociology can be used to intervene in the natural evolution of society for the improvement of society. Social Darwinism theorized that society was best left alone to follow its natural evolutionary course. Positivism is a system of thought in which scientific observation is considered the highest form of knowledge.

9. B, 13 Durkheim's research on solidarity reflects the principles of functionalist theory, which emphasizes the importance of integration, cohesion, and social order.

10. B, 15 Weber advocated that sociologists develop *verstehen*, or the ability to understand human behavior from the perspective of the people engaged in it. Social solidarity is a term used by Durkheim to refer to the bonds that link members to a group. Debunking refers to questioning taken-for-granted assumptions about social life.

11. A, 6 Berger referred to debunking as the process of questioning the taken-for-granted assumptions of social life. Sociological research may debunk commonly held, but inaccurate, ideas about human behavior and social life, in one's own society or others.

12. D, 15 Social Darwinism supported a "hands-off" approach to society, which would allow it to follow its "natural" evolutionary course. This perspective adopts ideas from Charles Darwin's work on biological adaptation in animal species.

13. B, 4 According to Mills, troubles result from problems originating in individual emotions or conditions, such as being fired for oversleeping. Issues arise from social conditions, such as layoffs due to economic restructuring.

14. C, 5 Issues such as mass layoffs and widespread unemployment affect large numbers of people and are rooted in the economic structure of society.

15. D, 16 Sociologists affiliated with the Chicago School, including W. I. Thomas, Florian Znaniecki, Robert Park, and Jane Addams, were interested in how society shaped the mind and identities of individuals. They used social settings as laboratories for human research to develop ways to solve social problems related to urbanization and industrialization.

16. D, 16 Jane Addams was an early sociologist of the Chicago school, a leader in the settlement house movement, and recipient of the Nobel Peace Prize in 1931.

17. B, 17 Merton identified the unintended consequences of behavior or social institutions as latent functions. The stated or intended goals are manifest functions.

18. B, 25 This participant observation study is an example of qualitative research, which is somewhat less structured and more interpretive than quantitative research, which relies more heavily on statistical analysis.

19. D, 19 Postmodernism is a theoretical perspective that suggests society is reflected in the words and images that people use to represent behavior and ideas.

20. B, 22 The independent variable in a hypothesis is the one that the researcher wants to test as the presumed cause of something else. The dependent variable is the one on which there is a presumed effect. In this example, gender is presumed to effect test performance.

21. D, 33 Rollins posed as a maid to conduct research on the experiences of Black domestic workers without revealing her identity to her employer or the other maids; thus, she did not get informed consent from the people being studied. Most sociologists do not believe that this did not constitute an ethical violation because her subjects were not at risk of being harmed as a result of the research.

22. C, 25 Surveys allow researchers to ask large numbers of people specific questions about many topics. Computers allow researchers to conduct sophisticated analyses in a relatively short amount of time. People may disguise their true opinions if they are concerned that their answers are not socially acceptable.

23. C, 27-8 Content analysis is a useful technique for examining the portrayal of groups in magazines and other media.

24. C, 29 Several statistical tools are defined in the "Statistics in Sociology" box. The mean is the average of all scores, the mode is the most frequently appearing score, and the median is the midpoint in a series of values. Unlike the mean, the median is not skewed by extreme values at either end.

25. D, 22 A hypothesis indicates the expected relationship between two variables that can be scientifically tested. It is usually formulated as an if-then statement.

True-False Questions

1. T, 17-8 Karl Marx believed that the economy was the central social institution around which other features of society were organized.

2. F, 13 In *How to Observe Manners and Morals*, Harriet Martineau discussed how to observe behavior when one is participant in the situation being studied.

3. F, 15 W. I. Thomas' saying is referred to as the "definition of the situation."

4. F, 16-7 Social Darwinism supported a "hands off" approach to society. Ward advocated social telesis, or the use of sociology for the improvement of society.

5. F, 11 The group of maps depicted in Map 1.1 indicate that the largest concentrations of Black Americans are in southern states such as Georgia and Alabama. Texas and California contain the largest concentrations of Hispanic people in the nation.

6. T, 12 Auguste Comte, who coined the term sociology, believed that sociology could discover the laws of social behavior just as science had discovered the laws of nature. Comte viewed sociology as the most highly evolved of the sciences.

7. T, 17 Merton elaborated on functionalist theory by identifying latent functions, or the unintended consequences of social behavior.

8. F, 21 The overall logic and strategy of a research project is called the research design.

9. T, 22 Quantitative data refers to numerical data that lends itself to statistical analysis.

10. F, 24-25 Deception is often used in controlled experiments that depend on respondents giving natural, spontaneous responses; participant observation may also be done without the knowledge of the people involved. Most sociologists would say these situations are acceptable as long as the research poses no risk of harm to the people being studied.

Fill-in-the-Blank Questions

1. facts, 14
2. validity, 23
3. random, 23
4. replication, 21
5. dependent, 22

Essay Questions

1. Troubles and Issues. C. Wright Mills' concept of the sociological imagination is discussed on p. 4-5 of the text. Students should demonstrate a clear understanding of the difference between a sociological perspective on public issues and copy with a personal trouble.

2. Diversity. The sociological concept of diversity is discussed on p. 9-10 in the text. Students should demonstrate a clear understanding of the sociological use of the term diversity to refer to the variety of group experiences that result from the social structure of society.

3. Statistical Mistakes. Statistics and the ways in which statistics can be misinterpreted and misused are discussed on p. 29 in the text. Student explanations and examples should demonstrate a clear understanding of basic statistical concepts and why their examples are misinterpretations or mistakes.

4. Research Design and Tools. The research process is described from p. 21-24, with a table of the process steps on p. 21. Students should demonstrate understanding of the research steps, as well as the basic components of research design. Research tools are described on p. 24-28. Students should demonstrate that they understand the appropriate uses of each of the three tools cited in this question.

5. Theoretical Frameworks. The three major theoretical frameworks used by sociologists are described on p. 16-19, with a table distinguishing the main features of each framework on p. 19. Student descriptions of how each framework would study school violence should demonstrate how the three frameworks differ in their understanding of the relationship of individuals to society, inequality, social order, and the sources of social change.

CULTURE

BRIEF CHAPTER OUTLINE

Defining Culture

Characteristics of Culture

Biology and Human Culture

The Elements of Culture

Language

Norms

Beliefs

Values

Cultural Diversity

Dominant Culture

Subcultures

Countercultures

Ethnocentrism

The Globalization of Culture

Popular Culture and the Media

The Influence of the Mass Media

Race, Gender, and Class in the Media

Theoretical Perspectives on Culture

Culture and Group Solidarity

Culture, Power, and Social Conflict

Symbolic Interaction and the Study of Culture

Cultural Change

Culture Lag

Sources of Cultural Change

Chapter Summary

CHAPTER FOCUS

This chapter identifies the elements of culture, discusses the influence of culture on diverse groups' experiences, describes how sociologists study culture, and introduces the sources and impact of cultural change.

QUESTIONS TO GUIDE YOUR READING

1. What is culture and what characteristics do all cultures share?

2. How does language shape culture and culture shape language?

3. How are social inequalities reflected in language?

4. How do norms govern situations and how do sanctions enforce norms?

5. How can cultural beliefs and values be sources of both cultural cohesion and social conflict?

6. What are subcultures and countercultures and what is their relation to the dominant culture?

7. What is ethnocentrism, and why is it potentially dangerous?

8. What is global culture, and what is the difference between popular and elite culture?

9. What is the influence of mass media, including their portrayal of race, gender, and class?

10. How do the major theoretical perspectives, including cultural studies, understand and study culture?

11. What is culture lag and culture shock?

12. What are the main sources of cultural change?

KEY TERMS

(defined at page number shown and in glossary)

beliefs 45

cultural capital 57

cultural hegemony 57

culture 36

culture shock 58

ethnocentrism 50

folkways 43

language 41

mass media 52

mores 43

norms 43

reflection hypothesis 54

social capital 57

subculture 48

KEY PEOPLE

(identified at page number shown)

CHAPTER OUTLINE

I. **DEFINING CULTURE**

The "Nacirema" article illustrates the importance of suspending judgments such as "strange" or "normal" when studying culture. Sociologists attempt to know a culture as an insider and to understand it as an outsider. **Culture**, the complex system of meaning and behavior that defines the way of life for a given group of people, includes customs, habits, dress, beliefs, values, knowledge, art, morals, language, and laws). **Material culture** consists of the objects created in a society, such as buildings, art, tools, toys, and print and broadcast media. **Non-material** culture includes the norms, laws, ideas, and beliefs of a group of people. Non-material culture is less tangible, but has a strong influence on behavior.

A. Characteristics of Culture
  1. *Culture is shared.* Culture is collectively experienced and agreed upon. The shared nature of culture makes society possible. Despite variations within the United States, certain symbols, language patterns, belief systems, and ways of thinking form a common American culture.
  2. *Culture is learned.* Even though the cultural beliefs and practices may seem perfectly natural to the members of the culture, they are learned through the formal and informal transmission of culture.
  3. *Culture is taken for granted.* Members of a culture seldom question their own culture; however, if a person becomes an outsider or establishes critical distance from typical cultural expectations, s/he may be able to examine the culture from a unique perspective. Although culture binds us together, lack of communication across cultures often has negative consequences.
  4. *Culture is symbolic.* **Symbols** are things or behaviors to which people give meaning, such as a flag or wedding band. Symbolic meanings guide behavior. The meaning attached to symbols

depends on the cultural context in which they appear, and can ignite controversy, as in the case of the confederate flag or Native American mascots of sports teams.

5. **Culture varies across time and place.** People develop cultural solutions to adapt to the challenges posed by their particular physical and social environments. Culture links the past and the present as it gives shape to human experience. **Cultural relativism** is the idea that something can be understood and judged only in relationship to the cultural context in which it appears. Practices accepted within certain cultures, such as burying or cremating the dead, may be viewed negatively by members of other cultures.

B. Biology and Human Culture

In addition to humans, some animal species, such as chimpanzees, develop culture. Although biological and environmental conditions place limitations on human development, cultural factors have an enormous influence on human life.

II. **THE ELEMENTS OF CULTURE**

A. Language

Language is a set of interrelated symbols and rules that provides a complex communication system. Language is fluid and dynamic, and makes the formation of human culture possible.

1. *Does Language Shape Culture?* The **Sapir-Whorf hypothesis** stated that language determines other aspects of culture because language provides the categories through which social reality is defined and constructed, forcing people to perceive the world in certain terms. Although critics do not agree that language is this deterministic, sociologists agree that culture and language influence each other. For example, concepts of time in the United States are strongly linked to the capitalist work ethic.

2. *Social Inequality in Culture.* Patterns of race, gender, and class inequality are reflected in language. Names of various racial and ethnic groups have been heavily debated, because what someone is called imposes an identity on that person. Power relationships between groups supply the social context for the connotations of language. Language can reproduce racist and sexist thinking, but by changing the language people use, we can alter social stereotypes and thereby change the way people think.

B. Norms
1. **Norms** are the specific cultural expectations for how to behave in certain situations. Norms exist to govern every situation, contributing to consistency and predictability in social interaction.
   a. Implicit norms, such as waiting in line rather than barging in front of people, need not be spelled out for people to understand them.
   b. Norms are explicit when the rules governing behaviors are written down or formally communicated.
2. **William Graham Sumner** (1906) identified two types of norms:
   a. **Folkways** are the general standards of behavior adhered to by a group, or the ordinary customs of different groups, such as fashion and etiquette.

  b. **Mores**, the strict norms that control moral and ethical behavior, are often upheld by **laws**, which are the written set of guidelines that define right and wrong in a society.

 3. **Social sanctions**, which are mechanisms of social control that enforce norms, may be imposed on people who violate norms.

  a. Negative sanctions may be mild or severe, ranging from ridicule to imprisonment and physical coercion.

  b. The strictest norms are **taboos**, which are those behaviors that bring the most serious sanctions.

  c. Sanctions also include rewards, such as praise and encouragement, which reinforce socially acceptable behavior.

 4. **Ethnomethodology** is a technique for studying human interaction that involves deliberately disrupting social norms and observing how individuals respond.

C. Beliefs

Beliefs are shared ideas held collectively by people within a given culture that form the basis for many norms and values. Beliefs provide a meaning system around which culture is organized, such as the belief in democracy in the United States.

D. Values

 1. **Values** are the abstract standards in a society or group that define ideal principles and provide a general outline for behavior by identifying what is desirable and morally correct.

 2. Values can be both a basis for cultural cohesion and a source of conflict, as illustrated by the Terri Schiavo case.

 3. Values guide the behavior of people in society, and norms reflect those underlying values. For example, the American Indian society, Kwakiutl, participates in a practice called *potlatch* that reflects its value of reciprocity, while *conspicuous consumption* characterizes dominant American culture.

III. CULTURAL DIVERSITY

The United States hosts enormous cultural diversity, where more than 11 percent of the population is foreign born, with most immigrants coming from Latin America and Asia, resulting in multiple languages and a variety of cultural forms. Jazz is one of the few musical forms indigenous to the U.S. Native American cultures have also enriched the culture of U.S. society.

A. Dominant Culture

As the culture of the most powerful group in society, the **dominant culture** receives the most support from social institutions and constitutes the major belief system in that society. A dominant culture need not be the culture of the majority of people.

B. Subcultures

Subcultures are the cultures of groups whose values, norms, and behavior are somewhat different from those of the dominant culture, but that share some elements of the dominant culture and coexist within it. For example, rap and hip hop, originally a subculture of young African Americans, has become mainstreamed but still represents the oppositional identities of Black and White youth for those groups who feel marginalized.

C. Countercultures

Countercultures are subcultures that reject the dominant cultural values, often for political or moral reasons, and may operate underground during a repressive regime. Some countercultures, like the contemporary militia movement, directly challenge the dominant political system.

D. Ethnocentrism

Ethnocentrism is the habit of seeing things only from the perspective of one's own group. An ethnocentric view prevents people from understanding the world as it is experienced by others, as an attitude of *cultural relativism* would. Although ethnocentrism may build group solidarity, it discourages understanding between groups, as in the case of *nationalism*, and may lead to terrorism, war, or even *genocide*.

E. The Globalization of Culture

From Disney films to McDonald's fast food items, the commercialized culture of the United States is marketed worldwide. The diffusion of a single culture throughout the world is known as **global culture**, which is increasingly marked by capitalist interests. Some, like Benjamin Barber, see many international conflicts as rooted in the struggle between this global, consumer-based, capitalist Western culture and traditional values of local communities.

IV. **POPULAR CULTURE AND THE MEDIA**

Popular culture includes the beliefs, practices, and objects that are part of everyday traditions, including mass-produced, mass-marketed media that are shared by large audiences. Popular culture is distinct from elite culture or "high culture," which is shared by only a select group who can afford to participate in it. Thus, cultural tastes and participation in the arts are socially structured, because familiarity with different cultural forms stems from patterns of historical exclusion, as well as integration into networks that provide information about certain cultural products. As popular culture is increasingly disseminated by the mass media, it is buttressed by the interests of big entertainment and information industries that profit from the cultural forms they produce.

A. The Influence of the Mass Media
   1. **Mass media** are those channels of communication that are available to wide segments of the population, including radio and television, which strongly shape public information and attitudes.
   2. Television is a powerful force for transmitting cultural values in the United States. For most Americans, leisure time is dominated by television, with the average person consuming 73 hours per week; 42% of homes are "constant television households."
   3. Media portray a very homogeneous view of culture, are ubiquitous, and have enormous power to shape public opinion and behavior.

4. Mass media, especially television, play a huge role in shaping people's perception and awareness of social issue. Fear of crime, for example, is directly related to the time spent watching television or listening to radio.

5. News is manufactured in a complex social process, involving commercial interests, the values of news producers, and their perceptions of what matters to the public.

B. Race, Sex, and Class in the Media

The mass media promote narrow definitions of who people are and what they can be. For example, the media communicate that only certain forms of beauty are culturally valued by portraying characters differently according to their age, gender, race, and class. Television shows, films, and music play a significant part in molding public consciousness, including upholding stereotypes about women as sexual objects. Even though African Americans and Hispanics watch more television than Whites do, they represent a small proportion of TV characters, in limited, stereotyped roles. Class stereotypes abound as well. Research indicates that these images do matter, impacting attitudes about relationships and self-image.

V. **THEORETICAL PERSPECTIVES ON CULTURE**

The **reflection hypothesis** contends that the mass media reflect the values of the general population by trying to appeal to the most broad-based audience; however, media portrayals can also influence the values of those people who see them.

A. Culture and Group Solidarity
   1. Sociologists such as Max Weber have studied the relationship of culture to other social institutions. Weber argued that the Protestant faith rested on cultural beliefs, such as the work ethic and a need to display material success as a sign of religious salvation, that were compatible with modern capitalism.
   2. Functionalist theorists believe that norms and values create social bonds that attach people to society, which provides coherence and stability. Putnam argues that a decline in civic participation in the U.S. has contributed to social disorder.

B. Culture, Power and Social Conflict
   1. Conflict theorists have analyzed culture as a source of power in society that is dominated by economic interests. A few powerful groups are viewed as the major producers and distributors of culture, a trend that is supported by corporate mergers in the media industry.
   2. Conflict theorists view culture as increasingly connected by economic monopolies, resulting in **cultural hegemony**, or an excessive concentration of cultural power that leads to the pervasive influence of one culture throughout society. This process creates a homogeneous mass culture that reduces political resistance to the dominant culture.
   3. Culture can also be a source of political resistance, as illustrated by the repatriation movement among American Indians.
   4. **Cultural capital** (or **social capital**) refers to those cultural resources that are socially designated as being worthy, such as knowledge of elite culture. Pierre Bourdieu argues that groups maintain their social status by appropriating culture.

C. Symbolic Interaction and the Study of Culture

Symbolic interaction theory analyzes behavior in terms of the meaning people give to it, noting that culture is produced through social relationships and in social groups. An interdisciplinary field influenced by *postmodernism* known as *cultural studies* has emerged that builds on the insights of this theory, directing researchers to view culture as a series of images that can be interpreted in multiple ways, depending on the viewpoint of the observer.

VI. CULTURAL CHANGE

Culture is dynamic and develops as people respond to changes in their physical and social environments, despite economic forces that support the status quo. Fast food culture, described by Eric Schlosser, illustrates how such change is hardly visible without taking a long-range view and questioning that which surrounds us.

A. Culture Lag

Culture lag refers to the delay in making cultural adjustments to changing social conditions. For example, culture lag is created when people's transportation habits do not change even though more efficient, lower-pollution forms of transit are available. When culture changes rapidly, or someone is suddenly thrust into a new cultural situation, the result can be **culture shock.**

B. Sources of Cultural Change

The main causes of cultural change are changes in societal conditions, cultural diffusion, innovation, and the imposition of cultural change by an outside group.

1. *Cultures change in response to changed conditions in society.*

   For example, cultural changes emerged as the Baby Boomers, the large cohort of children born after World War II, reached adulthood and became more affluent.

2. *Cultures change through cultural diffusion*

   Cultural diffusion is the transmission of cultural elements from one society or culture to another, such as the diffusion of rap music from inner city African American neighborhoods into White youth culture.

3. *Cultures change as the result of innovation.*

   The discovery and application of new knowledge, including inventions and technological innovations such as the microwave and the personal computer, have led to dramatic changes in lifestyles.

4. *Cultural change can be imposed*, as when a powerful group imposes a new culture on a society or manipulates the culture of a group as a way of exerting social control. Cultural expression can also be a form of political protest in which suppressed groups increase solidarity among members and attempt to establish a more powerful identity in the society. *Nationalist movements* are an example.

INTERNET EXERCISES

1. *Media Watch* (www.mediawatch.com/) provides provocative commentary on various depictions of racism, sexism, and violence in the media. Have students review current news items and action events. The site also offers numerous links to other sites.

2. *Media Channel* (www.mediachannel.org/) is a global Internet network of 1107 organizations focused on media issues, dedicated to promoting democratic media. Have students select an issue that interests them and report on how the issue is framed, what organizations are involved, and what global connections exist.

3. Other media websites that could be useful in exploring media issues include the following: *Media Alliance* (www.media-alliance.org/) is dedicated to promoting media excellence, ethics, diversity and accountability in the interests of peace, justice, and social responsibility. ACME: Action Coalition for Media Education (www.acmecoalition.org/) includes sections on media literacy and reform, research, independent media, and resources for parents, activists, and students. MEF: Media Education Foundation (www.mediaed.org/) produces extensive educational videos. Its website includes study guides, handouts and downloads, as well as articles on various media-related topics. *Center for Digital Democracy* (www.democraticmedia.org/) is in the forefront of media analysis and activism for democratic media. *Moving Ideas* (www.movingideas.org/), a project of The American Prospect, has an informative and richly linked site on Media and Culture, including a section on Media Consolidation.

From all of these sites, follow the links on particular issues of interest.

INFOTRAC EXERCISES

Conduct a keyword search using InfoTrac College Edition to extend the discussion in DOING SOCIOLOGICAL RESEARCH: *Tattoos: Status Risk or Status Symbol?* (p. 48-49)

1. **Keyword: Body art.** Searching with this term leads primarily to articles about safety and health risks, rates of piercing and tattooing in the population, and historical and cross-cultural significance of the practice.

2. **Keyword: Tattooing.** This keyword leads to articles on regulations, health risks, high-risk behaviors among adolescents, use of the practice by criminals and youth offenders, and many articles on psychological motivations, similar to the search with body piercing.

3. **Keyword: Body piercing.** In addition to a focus on regulations and health and safety concerns, body piercing appears in articles on Title VII (workplace accommodation of religious practice and belief), body art as a search for the self, and the psychological motivations of adolescent users.

PRACTICE TEST

Multiple Choice Questions

1. Certain elements of Black urban street culture, such as rap music, have been fully integrated into the dominant American culture. This is an example of cultural _____.
   a. lag
   b. diffusion
   c. innovation
   d. oppression

2. The norms, laws, customs, ideas, and beliefs of a group of people constitute its _____ culture.
   a. global
   b. counter
   c. material
   d. nonmaterial

3. Which of the following statements about culture is **false**?
   a. Culture is learned by direct instruction and observation.
   b. Culture is innate, or part of the basic instinctive patterns of human beings.
   c. Cultural values are represented by symbols to which people attribute meaning.
   d. Culture is fluid, so it changes in response to new environmental and social conditions.

4. The idea that something can be understood and judged only in relationship to the cultural context in which it appears is cultural _____.
   a. hegemony
   b. relativism
   c. shock
   d. lag

5. Which theory argues that culture serves the interests of powerful groups in society and is increasingly connected by economic monopolies?
   a. symbolic interaction
   b. functionalist theory
   c. solidarity theory
   d. conflict theory

6. Whereas earlier immigrants to the U.S. were predominantly from _____, the majority of current immigrants to the U.S. come from Latin America and from _____.
   a. Asia; Europe
   b. Europe; Africa
   c. North America; Asia
   d. Europe; Asia

7. The Sapir-Whorf hypothesis argues that language _____.
   a. determines other aspects of culture because it provides the categories through which social reality is constructed
   b. results from people trying to find ways to express new thoughts and perceptions
   c. is not necessary for acquiring the other social skills needed to participate in social life
   d. none of these choices

8. The feeling of disorientation when one encounters a new or rapidly changed cultural situation is called _____.
   a. culture lag
   b. jet lag
   c. culture shock
   d. cultural distress

9. At the grocery store, people are expected to wait in line, rather than barge in front of other people, to pay for the items they are purchasing. This is an example of a(n) _____.
   a. core value
   b. sacred more
   c. implicit norm
   d. explicit norm

10. William Graham Sumner called food preparation techniques and standards of etiquette _____.
    a. laws
    b. mores
    c. folkways
    d. sanctions

11. There is a strict norm against murder in the United States, and this morally offensive behavior may be severely punished. Thus, norms against killing are _____.
    a. mores
    b. beliefs
    c. folkways
    d. sanctions

12. Which of the following statements about sanctions is (are) true?
    a. Violations of mores carry stricter sentences than violations of folkways.
    b. Sanctions are mechanisms of social control that enforce norms.
    c. Sanctions involve both punishments and rewards.
    d. all of these choices

13. The technique for studying human interaction by deliberately disrupting social norms and observing how individuals respond is known as _____.
    a. postmodernism
    b. cultural studies
    c. etiquette analysis
    d. ethnomethodology

14. A shared idea held collectively by people within a given culture, such as cherishing democracy in the United States, is a _____.
    a. more
    b. norm
    c. value
    d. belief

15. An abstract standard that defines ideal principles within a given society, such as the primacy of individual freedom in the United States, is a _____.
    a. more
    b. norm
    c. value
    d. belief

16. Which of the following statements about ethnocentrism is **false**?
    a. Ethnocentrism discourages intercultural or intergroup understanding.
    b. Ethnocentrism can build group solidarity.
    c. Ethnocentrism in its extreme form can lead to genocide, or mass killing of people based on their membership in a particular group.
    d. Ethnocentrism, by definition, is not practiced by racial or ethnic minorities.

17. The greatest concentration of people speaking a language other than English at home is located in which region of the United States?
    a. Midwest
    b. Northeast
    c. Southwest
    d. Northwest

18. Which perspective views culture as a changing system that is socially constructed through the activities of social groups?
    a. conflict theory
    b. functionalist theory
    c. globalization theory
    d. symbolic interaction theory

19. The principle that suggests the mass media reflect the values of the general population is known as the _____.
    a. looking glass principle
    b. globalization process
    c. reflection hypothesis
    d. hegemonic hypothesis

20. Although we have the technology to develop more efficient, less polluting public transit systems, people's personal transportation habits have been difficult to change. This situation reflects the problem of cultural _____.
    a. lag
    b. diffusion
    c. ignorance
    d. oppression

True-False Questions

1. Sociologists agree that using ethnocentrism is the best way to fully understand and appreciate a culture other than their own.

   TRUE or FALSE

2. Material culture consists of the objects created by a given society, such as buildings, art, tools, and toys.

   TRUE or FALSE

3. Because culture is so critical to human survival, it is a stable, permanent system that rarely changes.

   TRUE or FALSE

4. According to the Sapir-Whorf hypothesis, social inequality in language is justified because women are naturally inferior to men, and language reflects that fact.

   TRUE or FALSE

5. Because participation in *elite culture* is so expensive, it is usually restricted to high status groups.

   TRUE or FALSE

6. *Nationalist movements* focus on celebrating the indigenous culture of an oppressed group as a mechanism for building group solidarity and resisting oppression.

   TRUE or FALSE

7. Subcultures may be based on a variety of shared understandings and experiences, including racial or ethnic heritage, religion, and musical preferences.

   TRUE or FALSE

Fill-in-the-Blank Questions

1. Cultural _____ refers to the pervasive, excessive influence of one homogeneous culture throughout a society.

2. The strictest norms in society are _____ .

3. According to _____ theory, culture provides coherence and stability in society and integrates people into groups.

4. Members of the modern militia movement, who reject dominant cultural values, physically isolate themselves from other Americans, and share a distinct style of life, represent a _____ .

5. Mass-produced, mass-marketed media that are shared by large audiences are a form of _____ culture, whereas expensive activities that are restricted to people of higher status are forms of _____ culture.

Essay Questions

1. Identify the four main sources of cultural change and provide an example of how each one has influenced behaviors or conditions in the United States.

2. Explain the statement, "Culture is symbolic." Give several examples of specific symbols and discuss how their meanings vary according to the social context in which they appear.

3. Discuss the issue of social inequality in language, noting how language reinforces the current power structure in the United States. Provide specific examples to support your answer.

4. Using an example of a major news story, explain how news is manufactured in a complex social process.

Solutions

PRACTICE TEST

Multiple Choice Questions

1. B, 59 Diffusion and innovation are both sources of cultural change. Diffusion refers to the transmission of cultural elements from one group to another. Innovation refers to the discovery of new knowledge, such as technological advances. Cultural lag refers to the delay in making cultural adjustments to new social conditions.

2. D, 36 Nonmaterial culture includes the norms, customs, laws, and beliefs of a group. Material culture includes the objects produced by the group, such as art.

3. B, 36-40 Culture is learned through socialization, which includes direct instruction and observation. Culture is fluid and dynamic, but not innate.

4. B, 40 Cultural relativism is the idea that something can be understood and judged only in relationship to its cultural context. Cultural hegemony refers to an excessive concentration of power that leads to the pervasive influence of one culture throughout society. Culture shock is a feeling of disorientation and alienation that can occur when a person encounters a new or changing situation.

5. D, 55-57 Conflict theory views culture as increasingly connected by economic monopolies. Functionalist theory views culture as a cohesive force in society. Symbolic interaction theory analyzes behavior in terms of the meaning people give to it.

6. D, 46-47 Map 2.2 indicates that the majority of earlier immigrants to the United States came from Europe, whereas the majority of recent immigrants have come from Latin America and Asia.

7. A, 41-42 The Sapir-Whorf hypothesis argues that language determines other aspects of culture because it provides the categories through which social reality is defined and constructed. Language itself determines what you think or perceive. Acquisition of language is necessary to full participation in society.

8. C, 58 Culture lag refers to the delay in cultural adjustments to changing social conditions. Culture shock is the feeling of disorientation when one encounters a new or rapidly change cultural situation.

9. C, 43 Values are abstract ideals. Mores are explicit norms governing moral behavior. Implicit norms are understood without being spelled out for people.

10. C, 43 Folkways are norms governing customary practices, such as table manners.

11. A, 43 Mores are strict norms governing moral behavior, such as murder.

12. D, 43-44 Social sanctions are mechanisms of social control that enforce norms. Sanctions include both rewards and punishments. The most severe sanctions are encoded in law and used for violations of mores.

13. D, 44-45 Ethnomethodology helps reveal the normal social order by deliberately disrupting patterns of human interaction. The field of cultural studies, which is based on postmodernism, examines how different people interpret the images that comprise culture.

14. D, 45 Beliefs are shared ideas held collectively by people within a given culture.

15. C, 45 Values are the abstract standards in a society or group that define ideal principles and provide a general outline for behavior.

16. D, 50 Ethnocentrism is the habit of seeing things only from the point of view of one's own group, which includes ignorance about other cultures. It can build group solidarity and discourages intercultural or intergroup understanding. In its extreme form, it can lead to genocide. Any group can be ethnocentric.

17. C, 44 Map 2.1 indicates that the Southwest region of the United States contains the highest percentage of people who speak a language other than English at home.

18. D, 55 Table 2.2 indicates that symbolic interaction theory focuses on the meanings people attribute to cultural symbols.

19. C, 54 The reflection hypothesis contends that the mass media reflect the values of the general population.

20. A, 58 Cultural lag occurs when there is a delay in adjustments to changes in society.

True-False Questions

1. F, 40 Sociologists advocate using cultural relativism to avoid judging cultural practices in different groups and societies.

2. T, 36 Material culture includes tangible objects such as art, while nonmaterial culture refers to the beliefs and norms of a society.

3. F, 40 Not only does culture vary from place to place, it also changes over time as people adapt to changes in the physical and social environment.

4. F, 41 According to the Sapir-Whorf hypothesis, language provides the categories through which social reality is defined and constructed by people. The hypothesis does not suggest that inequality among social groups is innate or natural.

5. T, 51 Popular culture is mass-produced and distributed, whereas elite culture is expensive. Thus, popular culture is available to the largest number of people, while elite culture is restricted to those of higher status.

6. T, 60 Nationalist movements, which identify a common culture as the basis for group solidarity, encourage members to celebrate their cultural heritage and challenge dominant cultural forms. Examples in the United States include the Black Power movement and the Latino movement, La Raza Unida.

7. T, 48-49 A subculture's values and norms of behavior differ from those of the dominant culture. Members tend to interact frequently with one another and share a common world view. Subcultures may be based a variety of cultural factors, including racial or ethnic heritage, religion, and musical preferences.

Fill-in-the-Blank Questions

1. hegemony, 57
2. taboos, 44
3. functionalist, 55
4. counterculture, 49
5. popular, 51; elite, 51

Essay Questions

1. Cultural Change. The four main sources of cultural change are discussed on p. 58-60 in the text. Students should demonstrate a clear understanding of each source of cultural change through their examples.

2. Symbols. The symbolic nature of culture is discussed on p. 38-39 in the text. Student examples should link the symbols to their cultural meaning.

3. Language and Inequality. Social inequality expressed in language is discussed on p. 42-43 in the text. Student examples should demonstrate understanding of the social inequalities to which the language refers.

4. News as Manufactured. The ways in which news is manufactured are discussed on p. 52-53 in the text. Student responses should demonstrate understanding of the ways in which news is socially constructed to reflect particular interests.

CHAPTER 3

SOCIALIZATION AND THE LIFE COURSE

BRIEF CHAPTER OUTLINE

The Socialization Process

The Nature-Nurture Controversy

Socialization as Social Control

Conformity and Individuality

The Consequences of Socialization

Agents of Socialization

The Family

The Media

Peers

Religion

Sports

Schools

Theories of Socialization

Psychoanalytic Theory

Social Learning Theory

Functionalism and Conflict Theory

Symbolic Interaction Theory

Growing Up in a Diverse Society

Aging and the Life Course

Childhood

Adolescence

Adulthood

Age and Aging

CHAPTER FOCUS

This chapter examines the process of identity development from various perspectives and in terms of diversity, describes socialization and the influence of socialization agents throughout the life course, and highlights issues related to age and resocialization.

QUESTIONS TO GUIDE YOUR READING

1. How does socialization establish identity and personality?

2. How do sociologists view the impact of nature and society on human identity?

3. Why do sociologists consider socialization a form of social control?

4. How do the family, media, peers, religion, sports, and schools influence socialization?

5. How do psychoanalytic, social learning, functional, conflict, and symbolic interaction theories explain the process of socialization and identity formation?

6. What is the influence of class, race, gender, and other factors of diversity on socialization?

7. What are the key tasks of socialization throughout the stages of the life course?

8. What is the impact of age stereotypes on a diverse elderly population?

9. What is the impact of ageism, age prejudice, and discrimination on life chances?

10. How does age stratification shape society and age cohort experience?

11. What are the roles of rites of passage that begin or end each life stage?

12. How do sociologists view resocialization, conversion, and so-called "brainwashing"?

KEY TERMS

(defined at page number shown and in glossary)

adult socialization 86

age discrimination 87

age stratification 87

anticipatory socialization 86

KEY PEOPLE

(identified at page number shown)

CHAPTER OUTLINE

I. The deciphering of the human genetic code raises the question of whether a human being can be created in a laboratory. Without society, however, what would humans be like? We have some idea from the experience of *feral children*. In 1970, Genie, a thirteen-year-old girl who had been kept in nearly complete isolation in her California home, was discovered. After intense language instruction and psychological treatment, she developed some verbal ability and showed progress in her mental and physical development, but she eventually moved to a home for mentally retarded adults.

II. **THE SOCIALIZATION PROCESS**

Socialization is the process through which people learn the expectations of society, including social **roles**, the expected behavior associated with a given status in society. Socialization is the basis for both **identity** and **personality**. The socialization process varies by race, ethnicity, gender, and social class. Socialization contributes to *internalization*, which occurs when behaviors and assumptions are learned so thoroughly that people no longer question them, but simply accept them as correct. People's lives are socially constructed; that is, the organization of society and the life outcomes of people within it are the result of social definitions and processes. For sociologists, what a person becomes is more a result of their social experiences than their innate, or inborn, traits.

A. The Nature-Nurture Controversy

From a sociological perspective, our lives are *socially constructed*. What a person becomes results more from social experiences than *innate* traits. Nature provides the stage, while society provides the full drama of what we become.

B. Socialization as Social Control

Socialization works as a subtle mechanism of **social control** because the socialization process brings individuals into conformity with dominant social expectations. When people successfully internalize their culture, they are likely to conform to social expectations. Deviating from cultural expectations can lead to ridicule as well as more coercive means of social control, including violence.

C. Conformity and Individuality

Despite the importance of social influences, human beings are individuals who interact with their environments in creative ways. Men and women who try to balance feminine and masculine characteristics enjoy greater mental health, while those who rigidly conform to their gender role may experience stress and other negative consequences.

D. The Consequences of Socialization

Socialization is a life-long process that affects how we think of ourselves and how we behave toward other people.

1. *Socialization establishes self-concepts*: identity is established through social experiences.
2. *Socialization creates the capacity for role-taking*: we are able to see ourselves as others see us through this reflective process.
3. *Socialization creates the tendency for people to act in acceptable ways*: we learn social expectations and this creates some predictability in human behavior.
4. *Socialization makes people bearers of culture*: we learn and internalize attitudes, beliefs, and behaviors and pass these cultural expectations on to others.

III. **AGENTS OF SOCIALIZATION**

Socialization agents are those who pass on social expectations. Socialization occurs in the context of social institutions, including the family, peers, the media, religion, sports, and schools.

A. The Family

Although the family is the first source of socialization for most people, families are quite diverse within and across cultures. For example, researchers found that Japanese mothers speak in ways that use objects as part of a ritual of social exchange in their interactions with their children, thereby emphasizing polite routines. American mothers, on the other hand, focus on labeling things for their children. These different styles of interaction are interpreted as reflecting the beliefs and practices of each culture.

B. The Media

In the U.S., the mass media—including television, print, film, music, video games, radio, and the Internet—have enormous influence on social values, societal images, desires, and relationships. Media violence reflects societal violence. Research finds that children imitate the aggressive behavior they see on television and in films, and that media violence encourages anti-social behavior and fear among children. Children are influenced not only by the images of televised and filmed violence alone, but also by the broader social context in which they live.

C. Peers

Peers are those people with whom you interact on equal terms, such as friends, fellow students, and coworkers. For children, peer culture is an important source of identity, where they learn concepts of self, gain social skills, and form values and attitudes. Members of minority groups often experience isolation and stress when they are "token" members of a dominant group, which may result in the formation of same-sex or same-race peer groups for support, social activities, and information sharing.

D. Religion

Religious instruction greatly contributes to the identities children construct for themselves and shapes the beliefs that guide adults in organizing their lives, including beliefs about moral development and behavior, the roles of men and women, sexuality, and child-rearing practices.

E. Sports
1. Through sports, men and women learn concepts of self and form ideas about gender differences. In interviews with male athletes, Michael Messner identified sports as very important to male identity. He reported that playing or watching sports is often the context in which men develop relationships with their fathers.
2. Although sports were less significant in the formation of women's identity in the past, women's participation in sports has increased. Traditionally, negative stereotypes of female athletes were a form of social control that reinforced traditional gender roles. Current research indicates that women who play sports develop a strong sense of bodily competence and self-confidence.

F. Schools
1. Research shows that teachers have different expectations for boys and girls and for students from different racial, ethnic, and social class backgrounds.
2. Negative appraisals are often self-fulfilling prophecies, because the expectations they create may become the basis for actual behavior, thereby affecting children's likelihood of success.
3. Schools have a *hidden curriculum* that is composed of informal and often subtle messages about social roles that are conveyed through classroom interaction and materials.
4. Gender separation heightens gender differences and greatly increases the significance of gender in children's interactions with each other.

5. Schools emphasize conformity to societal needs, such as respect for authority and punctuality; however, students internalize these lessons differently.

IV. **THEORIES OF SOCIALIZATION**

Several different theoretical perspectives have been used to explain the process of development. Each theory relies on unique assumptions about the effects of socialization on individual identity.

A. Psychoanalytic Theory
    1. **Psychoanalytic theory**, rooted in the work of **Sigmund Freud** (1856-1939), argues that the unconscious mind shapes human behavior. Psychoanalysis is used to discover the causes of psychological problems that exist deep within a patient's mind.
    2. This theory suggests the human psyche is comprised of three parts.
        a. The **id** consists of deep drives and impulses, such as sexuality.
        b. The **superego** is the dimension of the self that represents the standards of society. Because social standards (superego) will always be in conflict with impulses (id), individuals develop defense mechanisms such as repression, avoidance, and denial.
        c. As the psychological component of common sense and reason, the **ego** balances the id and the superego.
    3. Freud's work is controversial and questions have been raised about the ability to generalize his findings beyond his small, unrepresentative group of clients.
    4. Psychoanalytic theory is a popular way to think about human behavior that views identity as relatively fixed at an early age and motivations for behavior as internal and mostly unconscious.
    5. Nancy Chodorow uses psychoanalytic theory to explain how gender shapes men's and women's personalities, through attachment to the primary caregiver and later individuation. Since children also identify with their same-sex parent, according to Chodorow, women tend to have personalities based on attachment and an orientation toward others and men, personalities based on greater detachment.

B. Social Learning Theory
    1. **Social learning theory** considers the formation of identity to be a learned response to social stimuli. Identity is viewed as the result of modeling oneself in response to reinforcement.
    2. **Jean Piaget** believed that socialization and imagination have critical roles in learning, and noted that the human mind organizes experiences into mental categories called *schema*.
    3. Piaget proposed that children go through four distinct stages of cognitive development.
        a. In the *sensorimotor stage*, children experience the world directly through the senses-- touch, taste, sight, and sound.
        b. In the *preoperational stage*, children begin to use language and other symbols and to see things as others might see them, but cannot yet think abstractly.
        c. In the *concrete operational stage*, children learn logical principles about the concrete world.
        d. In the *formal operational stage*, children are able to think abstractly and imagine alternatives to their reality.
    4. Piaget conceptualized humans as actively creating their mental and social worlds, noting that people's behavior can be changed by altering their environments.

5. Lawrence Kohlberg elaborated on Piaget's work by developing a three stage theory of moral development.
   a. *Preconventional stage*: young children judge right and wrong in simple terms of obedience and punishment, based on their own needs and feelings.
   b. *Conventional stage*: adolescents develop moral judgment in terms of cultural norms, especially social acceptance and following authority.
   c. *Postconventional stage*: people are able to consider abstract ethical questions, demonstrating maturity in moral reasoning.
6. Kohlberg argued that men, who are more concerned with authority, reach a higher level of moral development than women, because women remain more concerned with feelings and social opinions.
7. Carol Gilligan challenged Kohlberg's theory, arguing that women's moral judgments were more contextual than those of men.

C. Functionalism and Conflict Theory

   Functionalists view socialization as one way society, by integrating people through internalization and reinforcing social consensus, maintains its stability. Conflict theorists focus on how group identity is shaped by patterns of inequality either in response to opportunity or by resistance to oppression.

D. Symbolic Interaction Theory
   1. According to symbolic interaction theory, people's actions are based on the meanings they attribute to things, and these meanings emerge through social interaction.
   2. For symbolic interactionists, the **self** is what we imagine we are, rather than an internal set of drives, instincts, and motives, and that people make conscious, meaningful adaptations to their social environment.
   3. Symbolic interactionists view socialization as a dynamic, ongoing process and the self as evolving over the life span. **Charles Horton Cooley** (1869-1929) and **George Herbert Mead** (1863-1931) viewed the self as developing in response to the expectations and judgments of other people in their social environments.
      a. Cooley developed the **looking-glass self** to explain how a person's conception of self develops through reflection about his/her relationships to others. This is a three-step process.
         i. We carefully note the reactions of others toward us.
         ii. We develop an understanding of how others judge us.
         iii. We develop feelings about ourselves based on the way we understand other people's perceptions of us.
      b. Mead believed that the basis of all social interaction is social **roles**, or sets of expectations that govern a person's relationships with other group members and society. He stated that the self has two dimensions: the active, creative, self-defining, unique part of the personality ("I") and the passive, conforming self that reacts to others ("me"). He also suggested that there are three stages of childhood socialization, based on the child's developing ability to engage in the process of **taking the role of the other**, or imagining oneself from someone else's point of view.

      i. **Imitation stage**: children only copy the behavior of those around them, without the ability to take on the role of the other.

     ii. **Play stage**: children begin to take on the roles of significant people in their environments. Of particular importance is taking on the role of **significant others**, or those with whom they have close relationships, such as their parents.

   iii. **Game stage**: the child becomes capable of taking on multiple roles at the same time; understands how people are related to each other and him or her; gains a more general, comprehensive view of the self; and acquires the **generalized other**, or the abstract composite of social roles and social expectations.

V. GROWING UP IN A DIVERSE SOCIETY

Socialization instills in us the values of the culture. It brings society into our self-definitions, our perceptions of others, and our understanding of the world. In a heterogeneous cultural system such as the United States, variation in social contexts creates vastly different social experiences. The socialization process is structured by social factors such as race, ethnicity, class, gender, religion, regional background, sexual preference, and age. For example, Annette Lareau's research on White and Black families from middle-class, working-class, and poor families demonstrated that social class is an important force shaping the socialization of young people.

VI. SOCIALIZATION AND THE LIFE CYCLE

Socialization begins at birth and continues throughout the lifespan. The term **life course** describes the connection between individuals, their roles, their experiences, and the social and historical context of life events. C. Wright Mills noted that personal biographies are linked to specific socio-historical periods.

A. Childhood
1. During childhood, socialization establishes one's initial identity and values, and the family is an especially influential agent of socialization.
2. Cultural experiences for children typically vary according to the sex and race of the child. For example, childhood play seems to encourage more rule-based, aggressive play for boys and more conversation in play for girls.
3. Ausdale and Feagin's research indicates that even preschool children use race and ethnicity to define themselves and others, often revealing awareness of negative racial attitudes (see *Children's Understandings of Race*, p. 82).
4. Although American culture defines children as "priceless," the image of childhood as a carefree time is inconsistent with the harsh realities of life for children who experience violence and poverty.

B. Adolescence
1. Adolescence developed as a distinct stage of the life course when formal education was extended to people of all social classes.
2. Erik Erikson stated that the central task of adolescence is the formation of a consistent identity. Conflict and confusion may arise as the adolescent moves between childhood and adulthood because adolescence lacks clear boundaries.
3. Patterns of adolescent socialization vary significantly by race, gender, and social class. Among the upper- and middle- classes, friendships tend to be based on shared activities and

interests, while working class youth tend to base friendship on loyalty and stability, with friendships determining activities.

C. Adulthood
1. **Adult socialization** involves learning behavior and attitudes appropriate to specific situations and roles, such as being a college student.
2. Events such as marrying, divorcing, beginning a career, and entering the military all transform an individual's identity and require the adult to adopt new roles.
3. Another part of learning new roles is **anticipatory socialization**, a process in which an individual learns, and perhaps rehearses, the expectations associated with a role that she or he expects to enter in the future.
4. In the transition from an old role to a new one, individuals may vacillate between their old and new identities For example, coming out, or openly identifying oneself as gay or lesbian, is an example of a process that typically occurs in stages and creates a new sense of self.
5. The transition to adulthood now takes longer than in the past, and social conditions make it difficult for many people to transition into adulthood by following the traditional path of finishing school, getting a job, marrying, and starting a family.

D. Age and Aging

The passage to old age is a difficult transition marked by biological processes, but always defined as a social phenomenon. **Age stereotypes**, reinforced through popular culture, differ for different groups.

1. **Age Prejudice and Discrimination.** Age prejudice relegates people of different age groups to lower status. While some forms of age discrimination are illegal, differential treatment based on age persists and **ageism** is manifested in the structure of institutions.
2. **Age Stratification.** Age stratification shapes the experiences of **age cohorts**, according to historical periods and societies. Drawn from functionalism, **disengagement** theory predicts that people gradually withdraw from participation and responsibility in society, providing an orderly transition from one generation to the next. Conflict theory focuses on the competition over scarce resources between age groups. Symbolic interactionism focuses on social definitions of aging and symbolic meanings attached to different age groups. The degree of satisfaction during old age depends on social support networks.

E. Rites of Passage
1. A **rite of passage** is a ceremony or ritual that marks the passage of an individual from one role to another. Rites of passage, including graduation ceremonies and religious affirmations, define and legitimize abrupt role changes that begin or end each stage of life. These events publicly announce the individual's new status.
2. Sociologists note that contemporary American society does not have a standardized, formal rite of passage marking the transition from childhood to adulthood, which contributes to the ambivalence and uncertainty of adolescence.

VII. RESOCIALIZATION

Resocialization is the process by which existing social roles are radically altered or replaced. Resocialization is especially likely when people enter institutions. It may involve degrading initiates physically and psychologically with the aim of breaking down or redefining their old identity.

A. <u>The Process of Conversion</u>. Conversion, whether or not extreme, involves a transformation of identity, and resocialization by changing beliefs and religious practices.

B. <u>The Brainwashing Debate</u>. Sociologists who have studied examples of *so-called brainwashing* view such conversions as *simply manifestations of the social influence people experience through interaction with others.*

    1. People susceptible to cult influence—primarily young adults who are socially isolated, drifting, and having difficulty performing in other areas—may choose to affiliate with cults voluntarily and many are able to leave on their own.

    2. Forcible confinement and physical torture can be instruments of extreme resocialization. Under conditions of severe captivity and deprivation, a captured person may come to identify with and become dependent upon the captor (the Stockholm Syndrome). This phenomenon may explain why some battered women do not leave their abusers or find it difficult to do so.

INTERNET EXERCISES

1. The Annie E. Casey Foundation (www.aecf.org/) works to build better futures for disadvantaged children and their families. Select an indicator from The Kids Count Data Book and search the latest year of national data and your state's data. Explore the graphs, maps and analysis of the indicator you selected, and summarize your findings, comparing your state with the national average. Next, review AECF projects and describe one project you think would work best to address the issue related to your indicator in your state.

2. The Children's Defense Fund (www.childrensdefense.org/) is a user-friendly, highly interactive site with information on programs for children, access to research, campaigns, and suggestions for action. Pick an issue or program that interests you and describe the research or action program of the CDF. Your description should include your critical assessment of the position and action program in terms of what you have learned about socialization.

3. National Institute on Aging (www.nih.gov/nia) The NIA is a rich source for the latest developments on health and aging. Select an issue that interests you and find fact sheets and reports on research in progress. For a broader perspective, go to the U.S. Administration on Aging (www.aoa.gov), and search for information on a topic like housing, poverty, civil rights, diversity, elders and families, or elder abuse. Review the latest annual edition of A Profile of Older Americans, which summarizes the current demographic data.

INFOTRAC EXERCISES

Conduct a keyword search using InfoTrac College Edition to extend the discussion of DOING SOCIOLOGICAL RESEARCH: Children's Understandings of Race (p. 82)

1. **Keyword: Racial socialization.** Searching with this term leads to articles on the influence of racial socialization by parents and the community on academic achievement, cognitive and behavioral competence, resilience, identity formation and stress management.

2. **Keyword: Learning or reducing racism.** Searching with either term will lead to limited but interesting studies on approaches to learning about racism and reducing awareness of race.

3. **Keyword: Children and prejudice.** This search leads to articles on the role of media, parents, religion and peers in learning or preventing prejudice in children.

4. **Keyword: Teachers and race prejudice.** Articles from this search look at teaching approaches that build on intergroup contact and mainstream teaching on race.

PRACTICE TEST

Multiple Choice Questions

1. The few rare individuals, who were raised in extreme isolation from other human beings and later found, sometimes referred to as feral children, are _____.
   a. usually able to reach a similar level of development as their peers after several years of intense language instruction
   b. never able to learn to speak the language of their culture because the capacity for learning language is only present during early childhood
   c. always able to emotionally recover from such severe neglect after they undergo intensive psychological treatment
   d. sometimes able to develop limited verbal ability and progress slightly in their physical and mental development after comprehensive training and treatment

2. Most people accept the norms of their society as correct because _____.
   a. norms are derived from the Bible and it would be blasphemous to question a sacred text
   b. people are innately motivated to desire similarity and seek conformity
   c. people internalize social expectations so well that they no longer question them
   d. most social norms are established by the government and people trust the government to act in their best interest

3. As shown in Map 3.2 The World's Children, based on U.S. Census Bureau data, the proportion of children under 15 years old as a percentage of the population of a given country tends to be higher in _____.
   a. the wealthier countries of North America and Western Europe, which can afford to have more children
   b. sub-Saharan and other African countries that are most economically disadvantaged and most overpopulated
   c. the Indian subcontinent
   d. South America and Australia

4. According to Erikson, the stage of the life course characterized by tension, contradiction, and the task of establishing one's identity is _____.
   a. childhood
   b. adolescence
   c. middle adulthood
   d. old age

5. Which of the following statements about socialization is (are) true?
   a. Socialization creates the tendency for people to act in socially acceptable ways.
   b. Socialization makes it difficult for individuals to see themselves as others see them.
   c. Socialization takes place during childhood and adolescence but is complete by adulthood.
   d. none of these choices

6. Carol Gilligan's research on moral development indicates that _____.
   a. men reach a higher standard of moral reasoning than women because men are more concerned with authority
   b. women reach a higher standard of moral reasoning than men because women are more concerned with relationships and other people's feelings
   c. women conceptualize morality in different terms than men, making more contextualized moral judgments
   d. there are no differences between women and men in the ways that they define morality and respond to moral dilemmas

7. According to Freud, the part of the personality that represents common sense and reason and helps balance the conflict between the other two components is the _____.
   a. superego
   b. ego
   c. me
   d. id

8. Which of the following statements about Freud's work on identity development is **true**?
   a. Freud's work has been criticized for being unrepresentative because he only wrote about women's development.
   b. Freud's work suggests that people make conscious, meaningful adaptations to their environments.
   c. Freud's work has been widely accepted because it was based on a representative sample that can be accurately generalized to a large population.
   d. Freud's work views identity as deeply rooted in the unconscious processes of the mind.

9. Symbolic interaction theory views identity as developing _____.
   a. through interaction and conscious, meaningful adaptation to one's environment
   b. from tensions between strong instinctual impulses and social standards
   c. from a passive, learned response to social stimuli such as reinforcement
   d. through the largely unconscious processes of attachment to, and separation from, one's caregivers

10. The relatively consistent pattern of behavior, feelings, predispositions, and beliefs in a given person is called _____.
    a. looking glass self
    b. personality
    c. identity
    d. superego

11. Which theory considers the formation of identity to be a learned response to social stimuli such as encouragement from others and rewards for desired behavior?
    a. Mead's game theory
    b. Piaget's social learning theory
    c. Freud's psychoanalytic theory
    d. Cooley's looking glass theory

12. According to Piaget, the preoperational stage of development is characterized by:
    a. experiencing the world directly through the senses of taste, touch, and sound.
    b. beginning to use language and other symbols but not thinking abstractly.
    c. thinking abstractly and imaging alternatives to one's own reality.
    d. learning logical principles about the concrete world.

13. According to Piaget, children are able to think abstractly and imagine alternatives to the reality in which they live in which of the following stages of development?
    a. formal operational
    b. concrete operational
    c. preoperational
    d. sensorimotor

14. Which of the following is **not** a component of the looking-glass self concept?
    a. Your perception of how you appear to other people
    b. Your beliefs about how other people perceive you
    c. Your feelings about how you are perceived by others
    d. Your beliefs about how you should treat other people

15. George Herbert Mead identified three stages of childhood. In which of the following stages do children typically begin to take on the roles of their significant others?
    a. Game
    b. Play
    c. Imitation
    d. Reflection

16. According to Mead, the abstract composite of social roles and social expectations that children acquire in the game stage is known as the _____.
    a. attachment process
    b. significant other
    c. resocialization effect
    d. generalized other

17. Through which source of socialization do boys and men typically form both an appropriate masculine identity and socially acceptable bonds with other men?
    a. Religion
    b. Sports
    c. Media
    d. Family

18. Tara, a graduate student, interacts regularly and on equal terms with the other graduate students in her program. The other students are Tara's _____.
    a. significant others
    b. generalized others
    c. models
    d. peers

19. According to symbolic interaction theory, the self is _____.
    a. an interior bundle of drives, instincts, and motives
    b. the same thing as personality
    c. what we imagine we are
    d. fully formed at birth

20. Barrie Thorne's study of school-aged children indicates that gender _____.
    a. becomes less relevant in the interactions of boys and girls when they are grouped together in common working groups
    b. is so central to the formation of identity that children should be grouped into boys-only and girls-only working groups to facilitate their development
    c. has a "rigid" character that cannot be changed to improve gender relationships between boys and girls, who both benefit from gender separation
    d. none of these choices

21. The theory that the elderly voluntarily withdraw from participation in society and are simultaneously relieved of responsibilities is _____.
    a. known as disengagement theory
    b. supported by cross-cultural and historical research on all known societies
    c. called generational balance
    d. known as age dysfunctionalism

22. Elizabeth is a high school senior who wants to attend college, so she visits several college campuses, observes how college students dress, and begins to imitate their actions. Elizabeth is engaged in _____.
    a. resocialization
    b. unsocialization
    c. optimistic socialization
    d. anticipatory socialization

23. What metaphor did Cooley use to describe how individuals form their identities?
    a. Undressing
    b. Playing games
    c. Looking in a mirror
    d. Constructing a building

24. When military recruits enter boot camp, their heads are shaved, they are given identical uniforms, and they must subordinate their identities to the group. This process is _____.
    a. resocialization
    b. unsocialization
    c. optimistic socialization
    d. anticipatory socialization

25. The phenomenon whereby a hostage identifies with his or her captor is known as the _____.
    a. Victimization Hypothesis
    b. Stockholm Syndrome
    c. Looking-Glass Self
    d. Imitation Stage

True-False Questions

1. According to Mead, the "I" is the component of self that is active, creative, and unique.

   TRUE or FALSE

2. During initiation into a hierarchical organization, the resocialization process promotes individuality through the personalized treatment of new members.

   TRUE or FALSE

3. Women who participate in organized sports tend to develop poor self-esteem because they are aware that they violate important cultural standards for femininity.

   TRUE or FALSE

4. Most people who join cults must be deprogrammed to be freed from the cult's influence, because they are rarely able to leave on their own.

   TRUE or FALSE

5. According to research conducted by Van Ausdale and Feagin, young children (ages three to five years old) do not understand the concepts of race and ethnicity well enough to use them as a basis for their concepts of themselves and other children.

   TRUE or FALSE

6. Psychoanalysis sees human behavior as directed and motivated by underlying psychic forces that are largely hidden from ordinary view.

   TRUE or FALSE

7. Age stereotypes are independent of gender, applying equally to men and women.

   TRUE or FALSE

Fill-in-the-Blank Questions

1. In the _____ stage of Mead's theory or socialization, children only copy the behavior of people around them and role taking is nonexistent.

2. According to Mead, the _____ other is the abstract composite of social roles and social expectations that children acquire in the game stage.

3. According to Erikson, the central task of _____ is the formation of a consistent identity.

4. A _____ is a far-reaching transformation of identity, often related to a change in political or religious beliefs.

5. According to Freud, the _____ is the dimension of self that represents the standards of society.

Essay Questions

1. Using Cooley's concept of the looking-glass self, explain why gay and lesbian people might develop low self-esteem. Use the same concept to explain how they might develop healthy self-esteem.

2. Discuss how the three perspectives—functionalism, conflict theory and symbolic interaction theory—explain age stratification.

3. Discuss how sociologists view "brainwashing."

4. Identify the four stages of the life course in the United States and describe the main activities and social expectations associated with each stage.

5. Explain what purpose rites of passage serve for both individuals and society, and discuss how adolescents in the United States could benefit from participating in a formal rite of passage.

Solutions

PRACTICE TEST

Multiple Choice Questions

1. D, 65-66 The few cases of feral children that have been scientifically documented have suffered severe intellectual and physical developmental problems and were unable to achieve a "normal" level of functioning.

2. C, 66 Socialization is the process through which people learn and internalize the expectations of society; thus, most people accept them as correct.

3. B, 85 As shown in Map 3.2, the proportion of children as a percentage of the population of a given country is highest in sub-Saharan and Central Africa, the most economically disadvantaged and most overpopulated countries in the world.

4. B, 84 According to Erikson, adolescence is a period characterized by conflict and confusion in which individuals have the task of establishing their identities.

5. A, 68 Socialization is a life-long process that creates the tendency for people to act in socially acceptable ways. It makes people bearers of culture and creates the capacity for role-taking, or seeing ourselves as others see us.

6. C, 78 Gilligan found that women conceptualize morality in different terms than men.

7. B, 76 According to Freud, the ego is the part of the personality that represents common sense and reason. It balances the conflict between the other two dimensions of self, the id and the superego.

8. D, 76 Freud's work has been criticized for relying on a small, unrepresentative sample of patients undergoing psychoanalysis. He argued that personality is relatively fixed at an early age and identity development is mainly the result of unconscious processes deeply rooted in the mind.

9. A, 79 Symbolic interaction theory views individuals as active participants in the ongoing development process. Psychoanalytic theory views personality as the result of tensions between unconscious processes and social standards. Social learning theory views development as a response to social stimuli. Object relations theory argues that young children's relationships with their caregivers determine adult development.

10. B, 66 Personality is the relatively consistent pattern of behavior, feelings, predispositions, and beliefs in a particular person. Identity is how one defines oneself. Superego is the dimension of the self that represents the standards of society. Looking-glass self is Cooley's model to explain how a person's conception of self arises through reflection about relationships to others.

11. B, 78 Social learning theory, associated with Piaget, considers identity formation to be a learned response to social stimuli. Looking-glass self is Cooley's concept for how the self emerges from relationships to others. Freud's psychoanalytic theory focused on unconscious mental processes. Mead's theory of role taking is based on symbolic interaction theory.

12. B, 78 According to Piaget, children experience the world through the senses during the sensorimotor stage. They use language but cannot think abstractly at the preoperational stage. At the concrete operational state, children learn logical principles about the concrete world.

13. A, 78 According to Piaget, children are able to think abstractly and imagine alternatives to the reality in which they live in the formal operational stage of development.

14. D, 79 In Cooley's model of the looking-glass self, individuals note the reactions of others toward them, understand how others view them, and develop feelings about themselves based on their perceived evaluations from others. Cooley's concept does not address how people feel about how they treat others.

15. B, 80 Mead identified three stages of childhood. In the imitation stage, children mimic the behavior of people around them. In the play stage, children take on the roles of significant people in their environment. In the game stage, children develop an awareness of community values and general social expectations.

16. D, 74 According to Mead, the abstract composite of social roles and social expectations that children acquire in the game stage is called the generalized other.

17. B, 73 Messner's research indicates that sports are very important in the development of masculine identity and the formation of male relationships.

18. D, 71 Individuals with whom a person interacts on equal terms, such as friends, fellow students, and co-workers, are peers.

19. C, 79 Symbolic interaction theorists see the self as what we imagine we are, and view it as developing throughout the life course, rather than as something set from birth or equivalent to personality.

20. A, 75 Thorne notes that gender becomes less relevant in boy-girl interactions when they are placed together in working groups. Because gender has a "fluid" character, relationships between boys and girls can be improved through conscious changes that discourage gender separation.

21. A, 88 Disengagement theory, drawn from functionalism, views reduced participation by the elderly as functional to society. Not all older people withdraw, and the status and engagement of elders varies historically and cross-culturally.

22. D, 86 Anticipatory socialization involves learning the expectations associated with a role one expects or hopes to enter in the future. Resocialization involves a dramatic transformation in one's identity.

23. C, 79 Cooley used the concept of the looking-glass self to compare the process of development to looking at one's reflection in a mirror.

24. A, 91 The process whereby existing social roles are radically altered or redefined is resocialization.

25. B, 92 Some hostages, prisoners of war, and battered women may identify with their captors, a phenomenon known as the Stockholm Syndrome.

True-False Questions

1. T, 79 According to Mead, the self has two dimensions. The "I" is the active, creative, self-defining part. The "me" is the passive, conforming part.

2. F, 91 Resocialization radically alters existing social roles and often involves breaking down members' individual identities to make them part of the group.

3. F, 73 Research indicates that women who participate in organized sports have high self-esteem and are confident and willing to take risks.

4. F, 92 Many people leave cults by themselves and do not require deprogramming.

5. F, 82 Preschool children use race and ethnicity to define themselves and others.

6. T, 77 Psychoanalysis sees human behavior as directed and motivated by underlying psychic forces.

7. F, 86 Gender is one of the most significant factors in age stereotypes.

Fill-in-the-Blank Questions

1. imitation, 80

2. generalized, 80

3. adolescence, 84

4. conversion, 91-92

5. superego, 76

Essay Questions

1. Looking-Glass Self. Cooley's concept of the looking-glass self in discussed in the text on p. 70. Student answers should demonstrate understanding of the looking-glass self and its relationship to self-esteem.

2. Age Stratification. Table 3.3 and the text p. 88-89 explain how the three perspectives approach the issue of age stratification. Student responses should demonstrate an understanding of age stratification as well as the three theoretical perspectives.

3. Brainwashing. So-called "brainwashing" is discussed in the text on p. 92, in the context of resocialization. Student responses should demonstrate why sociologists reject the concept of brainwashing and how they explain the conversion process.

4. Life Course. The four stages of the life course as they are experienced in the United States are discussed on p. 83-87. Student responses should demonstrate an understanding of the main activities and social expectations associated with each stage: childhood, adolescence, adulthood, and old age.

5. Rites of Passage. Rites of passage are discussed in the text on p. 90. Student responses should demonstrate the purposes rights of passage serve for both individuals and society, as well as the absence of a clear rite of passage for adolescents in the United States and the purpose that such a rite of passage might serve.

CHAPTER 4

SOCIAL INTERACTION AND SOCIAL STRUCTURE

BRIEF CHAPTER OUTLINE

What is Society?

From Groups to Institutions: Microanalysis and Macroanalysis

Groups

Statuses

Roles

Theories About Analyzing Social Interaction

The Social Construction of Reality

Ethnomethodology

Impression Management and Dramaturgy

Social Exchange and Game Theory

Interaction in Cyberspace

Forms of Nonverbal Communication

Touch

Paralinguistic Communication

Body Language

Use of Personal Space

Interpersonal Attraction and the Formation of Pairs

Proximity

Mere Exposure Effect

Perceived Physical Attractiveness

Similarity

Social Institutions and Social Structure

Social Institutions

CHAPTER FOCUS

This chapter examines the components and types of society, analyzes social interaction, communication, and social structure, and explores what holds societies together

QUESTIONS TO GUIDE YOUR READING

1. How is society both a system of social interaction and *sui generis*?

2. What is social organization and how are groups, statuses, and roles related to it?

3. How do the theories and perspectives of social construction of reality, ethnomethodology, impression management and dramaturgy, social exchange and game theory analyze social interaction?

4. How does cyberspace interaction differ from face-to-face interaction?

5. What are the four types of nonverbal communication and what role do they play in social interaction?

6. Which social factors are most important in determining personal attraction and the formation of pairs?

7. What are the major institutions in society and what are their primary functions?

8. How is social structure observable?

9. What holds societies together?

10. What is the difference between mechanical and organic solidarity, and between *gemeinschaft* and *gesellschaft*?

11. What distinguishes the six types of society, and what does it mean that the United States is suspended between industrial and post-industrial society?

KEY TERMS

(defined at page number shown and in glossary)

achieved status 99

collective consciousness 114

division of labor 115

gemeinschaft 115

group 99

imprinting 111

master status 100

microanalysis 98

paralinguistic communication 108

preindustrial society 116

role 100

role modeling 100

role strain 101

social interaction 98

social structure 114

status 99

status set 99

ascribed status 99

cyberspace interaction 106

ethnomethodology 103

gesellschaft 115

impression management 104

macroanalysis 98

mechanical solidarity 115

organic solidarity 115

postindustrial society 118

proxemic communication 109

role conflict 100

role set 100

social institution 113

social organization 98

society 98

status inconsistency 99

CHAPTER OUTLINE

I. **WHAT IS SOCIETY?**

Human **society** is a system of social interaction that includes both culture and social organization. Culture refers to people's general way of life, including norms, customs, beliefs, and language. Members of a society view themselves as distinct from other societies, maintain social ties through interaction, and have a high degree of interdependence. **Social interaction** refers to meaningful behavior between two or more people. Interaction involves communication, or the conveyance of information to other people. Durkheim viewed society as an organism, or something composed of different parts that work together to create a unique whole that is more than the sum of its parts (*sui generis*).

A. Microanalysis and Macroanalysis
   1. **Microanalysis** refers to the technique sociologists use to investigate the microlevel of society, or the patterns of social interaction that are relatively small, less complex, and less differentiated. For example, this approach could be used to study interpersonal attraction and the formation of friendships.
   2. **Macroanalysis** refers to the technique sociologists use to comprehend society as a whole, including how it is organized and how it changes. Macrolevel sociology investigates patterns of social interaction that are vast, complex, and highly differentiated. For example, this approach is useful for studying social problems such as poverty.

3. **Social organization** describes the order established in social groups. This order brings regularity and predictability to human behavior.

B. Groups
   1. Sociologists define a **group** as a collection of individuals who interact and communicate with each other, share goals and norms, and have a subjective awareness of themselves as a distinct social unit.
   2. In sociological terms, not all social units are groups.
      a. *Social categories* are people who are categorized together based on one or more shared characteristic, such as teenagers or teachers.
      b. *Audiences* are comprised of all of the people who are simultaneously watching the same program or performance.
      c. *Formal organizations* are highly structured social groupings that form to pursue a set of shared goals, including formal associations, such as the PTA, and bureaucracies, such as business corporations.

C. Statuses
   1. **Status** is an established position or rank in a social structure that carries with it a degree of prestige, or social value, such as a person's occupation.
   2. A **status set** is the complete set of statuses occupied by a person at a given time. Each status is associated with a different level of prestige.
   3. **Status inconsistency** occurs when a person's several statuses are associated with significantly different amounts of prestige. For example, recent immigrants who were professionals in their home countries but are forced into jobs with little status in the United States experience status inconsistency.
   4. There are two ways by which a person receives a particular status.
      a. **Achieved statuses** are those attained by independent effort, such as occupational and educational statuses.
      b. **Ascribed statuses** are those automatically assigned to a person at birth, such as race. These statuses are sometimes ambiguous, as in the case of biracial individuals. Although biological sex is ascribed, gender is a social construct, because regardless of genetic characteristics, gender appropriate behavior is learned, not innate.
      c. Class status includes both ascribed and achieved components. For example, upper-class membership is more likely if one's parents are wealthy (ascribed status) *and* if one works in a high-paying occupation (achieved status).
   5. A **master status** is a dominant status for an individual that overrides all other features of a person's identity. It may be imposed by others (e.g. criminal) or voluntarily chosen by the individual (e.g. mother). Because master statuses override other identities, they may be the basis for stereotypes, as in the case of someone who has a disability.

D. Roles
   1. A **role** is the expected behavior or collection of expectations associated with a particular status. A **role set** includes all of the roles occupied by a person at a given time.

2. One often learns roles through **role modeling**, a process by which one imitates the behavior of an admired person.

3. **Role conflict** occurs when two or more roles have contradictory expectations. For example, Arlie Hochschild identified the problem of the "second shift," which refers to women who are employed outside of the household but still expected to fulfill traditional expectations at home.

4. **Role strain** is a condition in which a single role brings conflicting expectations. For example, first-generation college students may experience role strain if their parents expect them to live at home and continue a traditional family role, while choosing the best college may require leaving home and thinking independently.

II. THEORIES ABOUT ANALYZING SOCIAL INTERACTION

Sociologists analyze social interaction through different theoretical frameworks and perspectives.

A. The Social Construction of Reality

1. The principle of the social construction of reality, central to symbolic interaction theory, argues that our perception of what is real is determined by the subjective meaning that we attribute to an experience. Things do not have their own intrinsic meaning; rather, people subjectively impose meaning on things.

2. As a result of *the social construction of reality*, we see what we want to see. In fact, people sometimes attribute certain meanings to things when it benefits them to perceive it that way, even if the perception seems to be contrary to fact.

3. W.I. Thomas coined the phrase *definition of the situation* to refer to the idea that *situations defined as real are real in their consequences*. Through the process of defining the situation, people adjust their attitudes and perceptions based on the context in which they find themselves. For example, researchers found that the way physicians define emergency room patients has important consequences for how the patients are treated. In one study, older patients were examined less thoroughly before being pronounced dead than were younger patients.

4. Understanding the social construction of reality allows us to gain insight into the social significance of race and gender. Race and gender have meaning because we choose to give them meaning, which may change our behavior toward others.

B. Ethnomethodology

Ethnomethodology is a technique for studying human interaction by deliberately disrupting social norms and observing how individuals try to restore normalcy. Because many norms that influence social behavior are not conscious, it is impossible to identify all of them simply by asking people to list them. For example, an ethnomethodologist might demand to pay more for a product than its listed price to reveal shopping norms.

C. Impression Management and Dramaturgy

1. **Impression management**, labeled by Erving Goffman, is a process by which people control how others perceive them by willfully attempting to manipulate other peoples' impressions of him or her. In this sense, impression management can be viewed as a type of *con game*.

2. Goffman's theory is sometimes referred to as the *dramaturgy model* of interaction because we present different "selves" to other people in different settings. This approach analyzes interaction by assuming that all participants are actors on a stage in the drama of everyday life.

3. Even when we define ourselves as individuals, our behavior is shaped by social forces. For example, a teacher's behavior when s/he returns graded papers is influenced by his/her relationship with the students and by the students' performance on the papers.

4. Social interaction is a perilous undertaking, as demonstrated by embarrassment, or a temporary challenge to one's identity, followed by an attempt to *restore face*.

D. <u>Social Exchange and Game Theory</u>

1. The *social exchange model* argues that social interactions are determined by the rewards or punishments that we receive from other people. If the reward for an interaction exceeds the punishments for it, a potential for social "profit" exists and the interaction is likely to continue. If the rewards are less than punishments, however, the interaction produces a social loss and will be less likely to continue.

a. Rewards include tangible gains, such as gifts, recognition, and money, as well as subtle rewards, such as smiles and hugs.

b. Punishments include both subtle gestures, such as a frown, and extreme behaviors, such as public humiliation, beating, and banishment.

2. Social exchange theory grew partly out of *game theory*, a mathematical and economic theory that predicts human interaction has the characteristics of a "game."

3. Exchange theory posits that racist and sexist stereotypes that are rewarded by one's group tend to persist, while those that are punished tend to change.

III. INTERACTION IN CYBERSPACE

When two or more persons share a virtual reality experience via communication and interaction with each other, they are engaged in **cyberspace interaction**. Cyberspace interaction differs from face-to-face interaction, because certain kinds of nonverbal communication are eliminated and one is free to become a different self or to engage in *impression management*. Cyberspace interaction has resulted in a new social order, including new social structure as well as a new culture—*cyberculture* or *virtual culture*. The benefits and harms of cyberspace interaction are the subject of research. Symbolic interaction theory predicts that in cyberspace interaction, *the reality of the situation grows out of the interaction process itself*.

IV. FORMS OF NONVERBAL COMMUNICATION

Social interaction includes both verbal and nonverbal communication. *Verbal interaction* consists of spoken and written language, whereas *nonverbal interaction* is conveyed by touch, tone of voice, gestures, body postures, eye contact, and facial expressions. The meanings of nonverbal communication are strongly dependent upon race, ethnicity, social class, and gender.

A. <u>Touch</u>

Tactile communication involves any conveyance of meaning through touch, whether positive (embracing) or negative (hitting). The meanings associated with tactile communication vary by cultural context and social factors such as gender. For example, as children, girls tend to be touched tenderly and protectively while boys are touched more roughly. As adults, women are more likely to touch as an expression of emotional support, whereas men touch more often to assert power or to express sexual interest.

B. Paralinguistic Communication
1. **Paralinguistic communication** is the component of communication that is conveyed by the pitch and loudness of the speaker's voice, its rhythm, emphasis, and frequency, and the frequency and lengths of hesitations.
2. The meaning of paralanguage varies by cultural context and ethnicity. For example, Japanese people regard silent periods during conversations as opportunities to collect their thoughts, while Americans avoid such periods with small talk. Americans often consider paralinguistics to be a small part of communication, while the Japanese consider them much more important.
3. *Nonverbal leakage* refers to an individual's emotions and feeling being revealed by paralinguistic slips despite the person's attempts to conceal them. For example, when a person is lying, the pitch of his or her voice is slightly higher than when that same person is telling the truth.

C. Body Language
1. Body language or *kinesic communication* involves gestures, facial expressions, and body language, which form a crucial part of nonverbal communication. Meanings conveyed by kinesis usually vary by cultural context, ethnicity, and gender. For example, avoiding eye contact is a sign of respect in some cultures, but making eye contact can be evaluated as a sign of sexual interest or hostility in other cultures.
2. Certain modes of kinesic communication, however, are identical across groups and different cultures. For example, the facial expressions for anger, happiness, sadness, and disgust are recognized in all cultures, as are many hand gestures, including "stop," "good-bye," and "OK."

D. Use of Personal Space
1. **Proxemic communication** refers to the amount of space present between interacting individuals. Generally, the more friendly a person feels toward someone, the closer he or she will stand.
2. According to E.T. Hall, each individual has a *proxemic bubble* that represents our personal 3-dimensional space. We feel threatened and may take evasive action when people we do not know enter our proxemic bubble. This largely unconscious process is illustrated by a typical person's behavior in an elevator.
3. Proxemic interaction varies by gender and ethnicity. The proxemic bubbles of different groups have different sizes, and people from different groups may experience difficulty when they interact if they are not aware of cultural differences in the use of personal space, as illustrated by Anderson's study, *Streetwise*.

V. INTERPERSONAL ATTRACTION AND THE FORMATION OF PAIRS

The formation of human pairings, including romantic couples and friendship groups, has a strong social structural component; that is, patterned by social forces. Humans have a strong need for *affiliation*, or a desire to be with other people, and women generally reveal this tendency more than men. The affiliation tendency is similar to **imprinting**, a phenomenon observed in newly hatched animals that attach themselves to the first living creature they encounter, regardless of species. In humans, infant attachment is a more complex and changeable process that is influenced more by social factors. Interpersonal attraction is a nonspecific, positive response toward another person that can be understood using sociological principles.

A. Proximity

Because we are more likely to meet and become attracted to people that we live or work near, proximity strongly affects relationship formation. In one study of friendships among recruits at a police academy, proximity in seating had a stronger effect on friendship formation than did all other factors, including race, socioeconomic background, and age.

B. Mere Exposure Effect

The *mere exposure effect* refers to the fact that the more you see someone, the more you like him or her. This effect even occurs when a person only sees someone in a photograph. The initial response of the viewer can determine how much liking will increase with exposure to additional photographs. Furthermore, "overexposure" can result when a photograph is seen too often and the viewer becomes "saturated."

C. Perceived Physical Attractiveness
    1. The attractions we feel toward people of either gender are based on our perceptions of their physical attractiveness. Perceived physical attractiveness is an important dimension of human interaction. For example, adults react more leniently to the bad behavior of an attractive child than to the same behavior of an unattractive child, and teachers evaluate cute children of either gender as "smarter" than physically unattractive children with identical academic records.
    2. Standards of attractiveness vary across cultures and among subcultures within a society, yet there is surprising agreement within a culture about who is attractive.
    3. Studies of dating patterns among college students show that the more attractive one is, the more likely one will be asked to go on a date; however, attractiveness predicts only the early stages of a relationship.

D. Similarity
    1. With few exceptions, people are attracted to others who are similar in socioeconomic status, race, ethnicity, religion, perceived personality traits, and general attitudes and opinions.
    2. The less similar a heterosexual relationship is with respect to race, social class, age, and educational aspirations, the quicker the relationship is likely to end.
    3. Although people tend to date within their own race, nationality, or ethnicity, many interracial couples enjoy long-lasting relationships. For interracial couples, similarities in other social characteristics tend to predict relationship duration.

VI. **SOCIAL INSTITUTIONS AND SOCIAL STRUCTURE**

A. Social Institutions

1. A **social institution** is an established and organized system of social behavior with a recognized purpose that develops to meet various needs in society.

2. Institutions cannot be directly observed, although their impact and structure can be studied. Specific schools are organizations where learning occurs, while at the broadest level, education is a social institution that includes all schools, as well as the norms, values, and beliefs that guide education at the societal level.

3. The major institutions in society include: family, education, work and the economy, politics (or state), religion, and health care. There are other institutions as well, such as the mass media and organized sports.

4. Functionalist theorists, who see societal needs as universal, have identified the purposes of institutions, although all societies do not fill these needs in the same way or through the same institution. In the United States:

    a. the *socialization of new members of the society* is a primary function of the family and education, although religious organizations and the mass media contribute to this function;

    b. the *production and distribution of goods and services* is the responsibility of the economic and political institutions;

    c. *replacement of the membership* is achieved through the family and other heterosexual pairings, although the government establishes policies that influence membership replacement;

    d. the *maintenance of stability and existence* is achieved through the government, law enforcement agencies, and the military; and

    e. *providing the members with an ultimate sense of purpose* involves virtually all institutions because commonly held purposes, values, and assumptions are evident in diverse institutions.

5. Conflict theorists argue that social institutions do not provide for all social members equally. Institutions affect individuals differently because they grant more power to some social groups than to others. Generally, the lower one's social class, the less one's political power, wealth, influence, and prestige.

B. Social Structure

1. **Social structure** refers to the organized pattern of social relationships and social institutions that together comprise society. Sociologists analyze social structure by examining the patterns in social life that reflect and produce social behavior.

2. Social class distinctions are an example of a social structure, because class shapes social interactions as well as the access that different groups have to social resources.

3. Social structures form invisible patterns that can be identified by using the sociological imagination. Marilyn Frye uses the metaphor of a birdcage to describe the concept of social structure. Just as a birdcage is a network of wires, society is a network of micro and macro structures that holds society together, sometimes oppressing certain groups.

VII. **WHAT HOLDS SOCIETY TOGETHER?**

Emile Durkheim argued that people in a society have a **collective consciousness**, or a body of common beliefs that that give people a sense of belonging and a feeling of moral obligation to its demands and values. Collective consciousness develops from participation in the common activities of social institutions. Durkheim, identified two different kinds of social solidarity.

A. Mechanical and Organic Solidarity
   1. **Mechanical solidarity** occurs when individuals play similar roles within the society. This form of solidarity characterized Native American groups prior to European conquest. Durkheim argued that collective consciousness was strongest in such societies; however, they are rare today due to the global trend of increasing interrelatedness.
   2. In societies with **organic solidarity** (or *contractual solidarity*), individuals play a great variety of different roles and social unity is based on role differentiation and shared interdependence, as in the United States and other industrialized societies. Such societies have a complex **division of labor**, or the systematic interrelatedness of different tasks. Within any division of labor, tasks become distinct from each other, yet they are woven together as a whole.
   3. Unlike societies characterized by organic solidarity, societies characterized by mechanical solidarity manifest tensions among competing groups, often due to divisions based on gender, race, and class.

B. Gemeinschaft and Gesellschaft
   1. *Gemeinschaft*, a German word meaning "community," refers to a society in which there is a sense of "we" feeling among members, a moderate division of labor, strong personal ties and family relationships, and a sense of personal loyalty. Social control is largely achieved through the sense of belongingness that members share.
   2. *Gesellschaft*, a German word for "society," refers to a society in which there is increasing importance on less intimate, more instrumental, secondary relationships. There is a reduced sense of personal loyalty to the total society, an elaborate division of labor, and a somewhat diminished role of the nuclear family. Social control is partly achieved through mechanisms external to the individual, such as the police.
   3. Social solidarity is weaker in a gesellschaft society than in a gemeinschaft society because gesellschaft societies are more likely to experience class conflicts and racial-ethnic divisions that reduce internal cohesiveness.
   4. Complexity and differentiation make the gesellschaft cohesive, whereas similarity and unity hold the gemeinschaft society together.

VIII. TYPES OF SOCIETIES: A GLOBAL VIEW

Societies are distinguished by different forms of social organization that evolve from both the relationship of the society to its environment, and from the processes that the society develops to meet basic human needs. Societies differ in many critical ways, including size, population, and resource base. Contemporary societies are increasingly global, with highly evolved systems of social differentiation and inequality, particularly along class, gender, racial, and ethnic lines. Sociologists distinguish six types of societies based on the complexity of their social structure, the amount of overall cultural accumulation, and the level of their technology [Table 4.1].

A. Preindustrial Societies

A preindustrial society is one that directly uses the land as a means of survival for society's members. There are four types of preindustrial societies.

1. *Foraging societies (hunting-gathering)* depend on hunting animals and gathering vegetation. Most are nomadic, such as the Pygmies of Central Africa.
2. *Pastoral societies*, which tend to be nomadic, depend on the domestication of animals in primarily desert areas, such as the Bedouins of Africa.
3. *Horticultural societies*, such as the Incas of Peru, are non-nomadic and use elaborate tools to cultivate the land.
4. *Agricultural societies* have large, complex economic systems and depend on technologically advanced, large-scale farming, as in the American south during the pre-Civil War period.

B. Industrial Societies
1. An *industrial society* uses machines and other advanced technologies to produce and distribute goods and services. The "industrial revolution," accompanied by the growth of science, brought advances in farming techniques and medical developments, for example, that led to the development of industrial societies.
2. Industrial societies rely upon a highly differentiated labor force and the intensive use of capital and technology. Large formal organizations (bureaucracies) and institutions with a high division of labor (economy and work, government and politics) are of critical importance in holding industrial societies together.
3. Industrial societies use a cash-based economy that pays wages for labor performed in factories, while household labor remains unpaid. This situation introduced the *family-wage economy*, in which families become dependent on wages to support themselves, but work within the family becomes increasingly devalued, and an increasing wage gap between men and women develops.
4. Industrial societies tend to be highly productive and have a large working class of industrial laborers. With industrialization, the population becomes increasingly urbanized, immigration is common, economic activities move outside of the family, and other institutions, such as schools, become increasingly important.
5. Although industrialization brought many benefits to the United States, it also produced some of the nation's most serious social problems, including pollution, widespread wage inequality and job dislocation, and urban crime and crowding.

C. Postindustrial Societies
1. **Postindustrial society** depends economically on the production and distribution of services, information, and knowledge.
2. The transition to a postindustrial society strongly influences the character of social institutions. For example, education and science become critically important, and workers without technical skills often find themselves in low-pay, unskilled work or permanent joblessness.
3. The United States is suspended between an industrial and a postindustrial society. Manufacturing jobs are still a segment of the labor force, but they are in decline. Most American workers are employed in the service sector of the economy, where they participate in the delivery of services and information rather than the production of material goods.

4. Postindustrial societies are increasingly dependent on the global economy as more goods are produced in economically dependent areas of the world for consumption in wealthier nations, resulting in greater global inequality.

INTERNET EXERCISES

1. Visit the website of Dr. John Suler's online book, The Psychology of Cyberspace (www.rider.edu/users/suler/psycyber/psycyber.html). Browse and find a topic that interests you, like gender swapping on the Internet. Based on your own experience of participating in chat-rooms or simply on reading sections of the book, write an analysis of the particular aspect of interaction in cyberspace you explored.

2. The website of the Center for Social Infomatics (www.slis.indiana.edu/CSI) has several interesting working papers on various social aspects of the Internet and communication technology. Read and write a brief summary of one paper that interests you. You might also want to explore Center-sponsored conferences and links to related sites.

INFOTRAC EXERCISES

Conduct a keyword search using InfoTrac College Edition to extend the discussion in DOING SOCIOLOGICAL RESEARCH: *Doing Hair, Doing Class* (p. 104-105).

1. **Keyword: Social interaction.** This search relates social interaction or deficits in social interaction to race, community, intergroup interactions, and mental illness.

2. **Keyword: Gender and social interaction.** This search leads to a variety of social interaction situations involving gender. You will want to narrow your search to your particular interest.

3. **Keyword: Social class.** This search leads to a range of articles in which social class is a variable.

4. **Keyword: Social status.** Depending on your interest, a search with social status or status attainment leads to a range of articles linking status to other variables.

5. **Keyword: Race and social interaction.** This search, like the search with gender, leads to a range of articles on social interaction involving race.

PRACTICE TEST

Multiple Choice Questions

1. Which model of social interaction suggests that racist and sexist behavior tends to persist when individuals are rewarded for such practices, yet those stereotypes are diminished when stereotypical attitudes result in punishment?
   a. social exchange model
   b. norm disruption model
   c. social conformity model
   d. construction of reality model

2. Sociologists refer to behavior involving communication between two or more people as social:
   a. interaction
   b. impression
   c. institution
   d. linguistics

3. A(n) _____ society uses machines and other technologies to produce and distribute goods, relies on a highly differentiated labor force, and uses a cash-based economy.
   a. horticultural
   b. agricultural
   c. industrial
   d. foraging

4. Which type of status do individuals gain through personal effort, such as educational attainment?
   a. achieved
   b. ascribed
   c. shared
   d. master

5. Which of the following statements about groups is (are) **true**?
   a. A group is a collection of individuals who interact and communicate with each other.
   b. Group members possess a subjective awareness of themselves as a distinct social unit.
   c. Group members share a common set of goals and norms.
   d. all of these choices

6. A highly structured social grouping that forms to pursue a set of shared goals is a(n) _____.
   a. social category
   b. social institution
   c. dispersed audience
   d. formal organization

7. Natasha is a recent immigrant who holds a college degree. She was a well-respected nurse in Russia, but she speaks limited English and is not licensed to practice nursing in the United States. She currently works as a maid, a job that carries little prestige. She is experiencing _____.
    a. status inconsistency
    b. status dysfunction
    c. role conflict
    d. role strain

8. Sociologists refer to those statuses that are automatically assigned to a person at birth, such as race, as _____.
    a. created
    b. ascribed
    c. achieved
    d. secondary

9. Mark was born with Cerebral Palsy, a physical disability that impedes his speech and mobility. Other people perceive Mark's disability as a dominant status that overrides all other features of his identity. Sociologists refer to Mark's disability as a(n) _____.
    a. status dysfunction
    b. achieved status
    c. master status
    d. organic role

10. Your boss asked you to work this weekend, you need to study for a test, and your mother expects you to come home from college for a family celebration. You are experiencing role _____.
    a. strain
    b. conflict
    c. disruption
    d. dysfunction

11. According to research on interpersonal attraction and couple formation, _____.
    a. there is considerable evidence that opposites attract
    b. the more you see someone, the more you like them
    c. characteristics such as race and class have relatively little influence on your selection of dating and marriage partners
    d. all of these choices

12. Which phrase did W.I. Thomas coin to refer to the idea that if people define things as real, then they are real in their consequences?
    a. Normative disruption
    b. Impression management
    c. Definition of the situation
    d. False construction of reality

13. According to ethnomethodologists:
    a. social interaction is determined primarily by the rewards and punishments we receive from other people.
    b. interpersonal attraction is determined primarily by biological and chemical processes located within the human nervous system.
    c. the most effective technique for identifying social norms is directly asking people to identify them.
    d. the most effective technique for identifying social norms is disrupting social settings.

14. Which of the following observations about cyberspace interaction is **false**?
    a. It has resulted in the formation of a new subculture in society.
    b. It eliminates the opportunity to develop intimate, long-term relationships.
    c. It provides greater anonymity than do most other forms of social interaction.
    d. It is supported by a set of beliefs and practices that encourage creating alternative identities.

15. The social positions, networks of relationships, and institutions that hold a society together and shape people's experiences make up _____.
    a. a status set
    b. an organization
    c. the social structure
    d. the division of labor

16. When Emily's mother asked her how she performed on her sociology test, she told her that she received a C even though she failed the test. Her mother knew that Emily was lying because the pitch and loudness of her voice betrayed her, reflecting the importance of _____ communication.
    a. paralinguistic
    b. proxemic
    c. kinesic
    d. tactile

17. Because Harry is sexually attracted to Joanne, he stands very close to her when they talk, which reflects his use of _____ communication to convey his feelings.
    a. paralinguistic
    b. proxemic
    c. kinesic
    d. tactile

18. Tactile communication refers to the _____.
    a. pitch and volume of our voices when we speak
    b. touching we do while we interact with other people
    c. distance we maintain between ourselves and other people
    d. facial expressions and gestures we exhibit while talking to other people

19. According to functionalist theorists, the purpose of the economic institution in the U.S. is to _____.
    a. replace members of society
    b. maintain stability and existence
    c. socialize new members of society
    d. produce and distribute goods and services

20. Joe notices that some boys he would like to be friends with frequently play basketball after school. Although he is not that interested in sports, Joe talks excitedly about the upcoming championship playoffs when he is with the boys so that they will invite him to join their group. According to Goffman, Joe is practicing _____.
    a. social deception
    b. normative disruption
    c. impression management
    d. interpersonal manipulation

True-False Questions

1. Microanalysis refers to the technique sociologists use to investigate patterns of social interaction that are relatively small, less complex, and less differentiated.

   TRUE or FALSE

2. Like individuals, social institutions can be directly observed by people who want to study them.

   TRUE or FALSE

3. Marilyn Frye compares the concept of social structure to a birdcage because she views both birdcages and society as havens that protect vulnerable groups from harm.

   TRUE or FALSE

4. A gesellschaft society is held together by organic solidarity.

   TRUE or FALSE

5. Research indicates that perceptions of physical attractiveness influence how people are evaluated and treated, with attractive children being viewed as "smarter" and more attractive adult defendants receiving lighter sentences for their crimes.

   TRUE or FALSE

6. Most foraging and pastoral societies, such as the Pygmies and Bedouins of Africa, are nomadic.

   TRUE or FALSE

7. Social solidarity is stronger in a gemeinschaft than in a gesellschaft society due to the greater homogeneity of the population in a gemeinschaft.

   TRUE or FALSE

8. Because one's proxemic bubble is determined entirely by culture, the proxemic interaction of men and women from the same culture is identical.

   TRUE or FALSE

9. The key driving force behind the development of the different types of society is the development of technology.

   TRUE or FALSE

10. The development of postindustrial societies promises universal prosperity.

    TRUE or FALSE

Fill-in-the-Blank Questions

1. Role _____ occurs when a single role consists of conflicting expectations.

2. When a person rolls his or her eyes to indicate disbelief or frowns to indicate disappointment, s/he is using _____ communication to convey information.

3. According to Durkheim, less complex, more homogeneous societies in which individuals play similar roles within the society are held together by _____ solidarity.

4. Erving Goffman's _____ model is based on the idea that members of society interact with each other in different settings as if they are actors on a stage.

5. In a(n) _____ society, a majority of workers are employed in the service sector, where they participate in producing distributing information and services, rather than tangible goods.

Essay Questions

1. Describe the basic principles of symbolic interaction theory and explain how Gilman used this theory to analyze status differences between hair stylists and their clients.

2. Compare and contrast gesellschaft and gemeinschaft societies in terms of their social organization, division of labor, degree of heterogeneity, and mechanisms of social control.

3. Identify four negative consequences of industrialization and the corresponding social problems that have developed in the United States since industrialization occurred.

4. Apply Goffman's principle of impression management to the phenomenon of cyberspace interaction and explain how the rituals of cyberculture contribute to the specific ways that individuals behave when engaged in cyberspace interaction.

Solutions

PRACTICE TEST

Multiple Choice Questions

1. A, 105 The social exchange model of social interaction suggests that when an interaction elicits approval, it is more likely to be repeated than an interaction that incites disapproval.

2. A, 98 Social interaction is behavior involving communication between two or more people. Social institution is an abstract concept referring to an established, organized system of social behavior with a recognized purpose in society.

3. C, 117 As indicated in Table 4.1, an industrial society uses machines and other complex technologies to produce and distribute goods, relies on a highly differentiated labor force, and uses a cash-based economy

4. A, 99 Ascribed statuses, such as race and biological sex, are present at birth. Achieved statuses are gained through individual effort, such as educational attainment.

5. D, 99 A group is a collection of individuals who interact and communicate with each other. Group members share goals and norms and possess a "we" feeling.

6. D, 98 Social organization brings regularity and predictability to human behavior. A formal organization is a highly structured social grouping that forms to pursue a set of shared goals. A social category refers to people who are categorized together based on a shared characteristic. An audience includes all of the people watching a program or performance. A social institution is an established and organized system of behavior with a recognized purpose.

7. A, 99 Status inconsistency occurs when several statuses held by one person are associated with significantly different amounts of prestige. Role conflict occurs when the expectations of multiple roles conflict, while role strain occurs when one role has conflicting expectations.

8. B, 99 Statuses that are automatically assigned to a person at birth, such as race, are ascribed. Achieved statuses are gained through individual effort.

9. C, 100 The dominant status for an individual, which overrides all other features of the person's identity, is a master status.

10. B, 100 Role conflict occurs when multiple roles have contradictory expectations, while role strain occurs when a single role includes conflicting expectations.

11. B, 113 Shared social characteristics such as race and class increase the likelihood of attraction and relationship formation.

12. C, 102 W. I. Thomas argued that situations defined as real are real in their consequences. This concept is known as the *definition of the situation*.

13. D, 103 Because many norms are unconscious, asking people to identify them is not a useful strategy. Using the technique of ethnomethodology to disrupt social situations can reveal norms.

14. B, 106-7 Cyberspace interaction occurs when people interact and communicate with each other electronically through the use of a personal computer. People can develop long-term, intimate relationships in cyberspace.

15. C, 99 Social structure consists of the social positions, networks of relationships, and social institutions that hold a society together. A status set is the complete set of statuses occupied by a person at a given time.

16. A, 108-9 Paralinguistic communication is conveyed by the pitch and loudness of the speaker's voice, as well as its rhythm, emphasis, and frequency.

17. B, 109-10 The amount of distance maintained between people engaged in social interaction. Refers to proxemic communication. People who are sexually attracted to each other stand especially close.

18. B, 108 Tactile communication refers to physical touch, including positive (hugging) and negative (hitting) forms of touching.

19. D, 113 Functionalist theorists argue that all social institutions fulfill specific purposes in society. The economy produces and distributes goods and services, the political institution maintains stability and order, the family replaces membership and socializes members, and religion provides members with a sense of purpose.

20. C, 104 Impression management is a process by which a person consciously manipulates other people's perceptions of him or her. Goffman examined this process with his dramaturgy model of social interaction.

True-False Questions

1. T, 98 Microanalysis refers to the technique that sociologists use to investigate patterns of social interaction that are relatively small and immediately visible.

2. F, 113 Social institutions cannot be directly observed, but their impact and structure can still be studied.

3. F, 114 Frye uses the metaphor of a birdcage to describe social structure as an oppressive network that confines and exploits some social groups.

4. T, 115 A complex, diverse gesellschaft is held together by organic solidarity, whereas a more homogeneous gemeinschaft is held together by mechanical solidarity.

5. T, 112 Adults tend to treat attractive children more leniently than they do unattractive children when they misbehave.

6. T, 116-17 Foraging and pastoral societies have subsistence economies and are nomadic.

7. T, 115 Social solidarity is stronger in gemeinschaft societies, where the population is more homogeneous. Greater diversity in gesellschaft societies can cause conflict and social divisions that undermine solidarity.

8. F, 110 Proxemic interaction varies strongly by cultural differences and by gender.

9. T, 116 Technology is the driving force that shapes social organization and the economy.

10. F, 118-9 The development of postindustrial societies produces prosperity for the wealthier nations and the more highly educated, and poverty and inequalities resulting from the global social structure that postindustrialism produces.

Fill-in-the-Blank Questions

1. strain, 101
2. kinesic, 109
3. mechanical, 115
4. dramaturgy, 104
5. postindustrial, 119

Essay Questions

1. Symbolic Interaction Theory. Gilman's "Doing Hair, Doing Class" is discussed on p. 104-105 in the text. Student responses should demonstrate and understanding of symbolic interaction theory applied to face-to-face interaction.

2. Gesellschaft and Gemeinschaft. Gesellschaft and gemeinschaft societies are discussed on p. 115 in the text. Student responses should demonstrate an understanding of the differences between these two types of society in terms of social organization, division of labor, degree of heterogeneity, and mechanisms of social control.

3. Industrialization. Industrialization is discussed on p. 118 in the text. Student responses should demonstrate understanding of the consequences of industrialization in general, and social problems related to industrialization that have arisen in the United States.

4. Cyberspace Interaction and Impression Management. Cyberspace interaction is discussed on p. 106, and impression management on p. 104 in the text. Student responses should demonstrate understanding of impression management and how it occurs in cyberspace interaction.

GROUPS AND ORGANIZATIONS

BRIEF CHAPTER OUTLINE

Types of Groups

Dyads and Triads: Group Size Effects

Primary and Secondary Groups

Reference Groups

In-Groups and Out-Groups

Social Networks

Social Influence

The Asch Conformity Experiment

The Milgram Obedience Studies

The Prisoners at *Abu Ghraib*: Research Predicts Reality?

Groupthink

Risky Shift

Formal Organizations and Bureaucracies

Types of Organizations

Bureaucracy

Bureaucracy's Other Face

Problems of Bureaucracies

The McDonaldization of Society

Diversity: Race, Gender, and Class in Organizations

Functional, Conflict, and Symbolic Interaction: Theoretical Perspectives

Chapter Summary

CHAPTER FOCUS

This chapter explores types of groups, social influence, and formal organizations, and examines the characteristics and problems of bureaucracies.

QUESTIONS TO GUIDE YOUR READING

1. What are the characteristics of dyads and triads, primary and secondary groups, reference groups, in-groups and out-groups, and social networks, and what is their impact on behavior and identity?

2. What was learned about social influence and group effects from the Asch conformity experiment, the Milgram obedience studies, and studies of groupthink and risky shift?

3. How does research on social influence predict the treatment of prisoners in *Abu Ghraib*?

4. What are the characteristics of normative, coercive, and utilitarian organizations, and total institutions?

5. What are the characteristics of the ideal type of bureaucracy?

6. What is bureaucracy's other face and why and how does it develop?

7. How do the problems of ritualism and alienation arise in bureaucracies?

8. What is the McDonaldization of society and how is it manifest in the organizations that shape daily life?

9. What is the glass ceiling, and how can it be addressed?

10. What are eufunctions and dysfunctions of bureaucracies, and how do functional, conflict, and symbolic interaction perspectives explain them?

KEY TERMS

(defined at page number shown and in glossary)

attribution error 130

bureaucracy 138

dyad 127

formal organization 137

group size effects 127

instrumental needs 129

reference group 129

secondary groups 128

total institution 138

triadic segregation 127

attribution theory 130

deindividuation 136

expressive needs 129

group 126

groupthink 135

primary group 127

risky shift 136

social network 131

triad 127

KEY PEOPLE

(identified at page number shown)

CHAPTER OUTLINE

I.   Juries are groups, and groups behave differently than individuals. Studies have shown that jury verdicts correlate not just with evidence, but also with jury composition. Because groups are influenced by social forces, sociological analyses of group decision-making processes can help predict trial outcomes.

II. **TYPES OF GROUPS**

A **group** is two or more individuals who interact with one another, share goals and norms, and have a subjective awareness as "we." Certain gatherings are not groups, but may be *social categories*, such as truck drivers, or *audiences*, such as persons watching a movie. Sociologists are able to identify characteristics that reliably predict trends in the behavior of specific groups, as well as the behavior of particular individuals within groups.

    A. <u>Dyads and Triads: Group Size Effects</u>
       1. **Georg Simmel** (1858-1918) investigated the effects of size on groups, noting that the difference between a **dyad** (two people) and a **triad** (three people) creates entirely different group dynamics.
       2. Simmel identified **group size effects**, noting that triads are *unstable* social groupings, whereas dyads are relatively stable, because triads tend to separate into a *coalition* of a dyad against the *isolate* in the process of **triadic segregation**. The isolate can then opt to initiate a coalition with one of the dyad members, according to the principle of *tertius gaudens* ("the third one gains").

    B. <u>Primary and Secondary Groups</u>
       1. **Charles Horton Cooley** (1864-1929) introduced the concept of a **primary group**, or a group consisting of intimate, face-to-face interaction and relatively long-lasting relationships, such as the family.
       2. **Secondary groups** are larger in membership, less intimate, and less permanent. Although they tend to be less significant in people's emotional lives, some secondary groups can take on the characteristics of primary groups, in times of crisis or high stress, as when a community experiences a natural disaster.
       3. Primary groups usually serve **expressive needs** and secondary groups generally serve **instrumental needs**. The difference between primary and secondary groups is less in why they form, however, than in how strongly the participants feel about each other and how dependent they are on the group for identity.

    C. <u>Reference Groups</u>

Reference groups are those to which you may or may not belong, but that you use as a standard for evaluating your values, attitudes, and behaviors. Imitation of reference groups can have both positive and negative effects. Identification with reference groups can strongly affect self-evaluation. For example, multicultural educational programs may help Black and Latino children develop higher self-esteem.

D. In-Groups and Out-Groups
1. **W. I. Thomas** (1931) noted that people like and trust *in-group* members, as with families and fraternities, whereas *out-group* members are often regarded with suspicion.
2. Attribution theory asserts that all people make judgments about others, called *dispositional attributions*.
3. Pettigrew's summary of research on attribution indicates that individuals commonly generate a significantly distorted perception of the motives and capabilities of other people's acts based on whether that person is an in-group or out-group member.
   a. These misperceptions are **attribution errors**, or errors made in attributing causes for people's behavior to their membership in a particular group.
   b. Attribution error has several dimensions, which tend to favor the in-group over the out-group.
      i. When they observe improper behavior by an out-group member, onlookers are likely to attribute the deviance to the *disposition*, or perceived "true nature," of the wrong doer. This nature is viewed as genetically determined.
      ii. When an in-group member exhibits the same behavior, the act is commonly perceived as a result of the *situation* of the wrong doer, rather than his or her disposition.
      iii. If an out-group member performs in an exceptional way, it is often attributed to good luck, or the individual is viewed as "exceptional" for his or her group.
      iv. An in-group member who performs in the same positive way is credited for having a valued disposition.

E. Social Networks
1. A **social network** is a set of links between individuals or between other social units, such as bureaucratic organizations or entire nations. Individual membership in several groups provides links between groups and groups overlap. Research indicates that people get jobs, especially high paying, prestigious jobs, via personal networks more often than through formal job listings, want ads, or placement agencies. Networks are influenced by race, class, and gender and include religious, fraternal, and occupational groups.
2. Research into the *small world problem* indicates that networks make the world much smaller than commonly thought. Taylor's study of Black national leaders revealed that they form a closely-knit network that is considerably more dense than that of the longer-established White leadership. When considering only personal acquaintances, not including indirect links, one-fifth of the entire national Black leadership is included.

III. **SOCIAL INFLUENCE**

The groups in which we participate exert tremendous influence on us. Despite what social psychologist Philip Zimbardo calls the *Not-Me Syndrome* in reference to in-group conformity, sociological experiments reveal a dramatic gulf between what people think they will do and what they actually do.

A. The Asch Conformity Experiment

Asch's experiment showed that even simple objective facts cannot withstand the distorting pressure of group influence. In replications of his study, one-third to one-half of the subjects make a judgment contrary to objective fact, yet in conformity with the group. The pressure to conform rises as the number of confederates increases.

B. The Milgram Obedience Studies

In Milgram's first study, 65% of the volunteer subjects administered what they thought was lethal voltage on the shock machine. Subsequent studies under different circumstances and using a variety of subjects—men and women, different racial and ethnic—produced similar results. Milgram described the dilemma revealed by his experiments as a conflict between conscience and authority, a dilemma that resonated with the world captivated by the trial of Adolf Eichmann in Jerusalem. As Hanna Arendt argued, we need to look into ourselves to understand that evil on a grand scale is banal.

C. The Prisoners at Abu Ghraib: Research Predicts Reality?

In the spring of 2004, it was revealed that U.S. soldiers, acting as military police guards in Abu Ghraib prison in Iraq, had engaged in severe torture, including sexual and severe physical abuse, of Iraqi prisoners of war. The guards claimed they were only following orders. As the studies of social influence would predict, *the causes of the solders' behavior lie not in their personalities, but in the social structure and group pressures of the situation.*

D. Groupthink
  1. I. L. Janis described **groupthink** as the common tendency for group members to reach a consensus opinion, even if that decision is stupid.
  2. In his investigation of presidential decisions, Janis found that the process of groupthink is more likely to occur under four conditions.
     a. Group members possess *an illusion of invulnerability*, believing that members are so talented that any group decision will succeed.
     b. Group members develop *a falsely negative impression of those who are antagonists to the group's plans*, which results in an underestimation of their strength and power.
     c. Groups *discourage dissenting opinions*, and view dissent as disloyalty, thereby prohibiting a full discussion of alternative strategies.
     d. Group members develop *an illusion of unanimity*, so that despite the personal reservations of some group members, there is a prevailing sense that the entire group is in complete agreement.

E. Risky Shift

Risky shift, also called polarization shift, is another group phenomenon that helps explain why the products of groupthink are frequently calamities. Risky shift is the tendency for groups to weigh risk differently than individuals. The risky shift is probably caused by group-size effect known as **deindividuation**, which is the sense that one's self has merged with the group.

IV. FORMAL ORGANIZATIONS AND BUREAUCRACIES

A **formal organization** is a large, secondary group, highly organized to accomplish a complex task and achieve goals in an efficient manner. Although an organization is ultimately a collection of people, organizations develop their own routine practices and cultures that may be reflected in symbols, values, rituals, and norms.

A. Types of Organizations
1. *Normative Organizations*. People voluntarily join *normative organizations* to receive personal satisfaction and pursue goals they personally consider worthwhile.
   a. Examples include service and charitable organizations such as the PTA, the National Organization for Women (NOW), and the NAACP (National Association for the Advancement of Colored People).
   b. Gender, class, race, and ethnicity structure who joins which voluntary organizations. Lower-income people cannot afford membership in many organizations. Due to their historical exclusion from White voluntary organizations, many racial-ethnic minority groups formed their own.
2. **Coercive Organizations**. Membership in *coercive organizations*, such as prisons and mental hospitals, is largely involuntary. Goffman described these organizations as **total institutions**, or organizations cut off from the rest of society where residents are subject to strict social control over all aspects of their lives.
3. **Utilitarian Organizations**. *Utilitarian organizations* are large organizations, either for profit (e.g., Microsoft) or nonprofit (e.g., colleges), that individuals voluntary join for specific purposes, such as monetary reward.

B. Bureaucracy

A **bureaucracy** is a type of formal organization characterized by an authority hierarchy, a clear division of labor, explicit rules, and impersonality. An example is the federal government. Max Weber identified the characteristics of the *ideal type bureaucracy*.

1. There is a **high degree of division of labor and specialization** with clearly defined responsibilities and privileges.
2. There is a **hierarchy of authority** that is often depicted in a complex organizational chart, which is a diagram identifying the position of each participant in the chain of command.
3. There are detailed **rules and regulations**, designed to handle virtually all situations and problems that govern the activities of a bureaucracy.
4. Establishing efficient **impersonal relationships** is important because social interaction in a bureaucracy is supposed to be guided by instrumental criteria, such as organizational rules, rather than by social or emotional criteria.
5. **Career ladders** are created within bureaucracies where job candidates are supposed to be hired and promoted on the basis of their qualifications.
6. **Efficiency** is important to coordinate the activities of a large number of people who are pursuing organizational goals.

C. Bureaucracy's Other Face

1. In addition to the formal structure of a bureaucracy, an *informal structure* exists, which includes social interactions in bureaucracies that ignore, change, or otherwise bypass the formal structure and rules of the organization.
2. *Bureaucracy's other face*, or the informally evolved culture that evolves over time as a reaction to the formality and impersonality of the bureaucracy, can be seen in the workplace subcultures that develop.
3. The informal culture may become exclusionary, thereby increasing some workers' feelings of isolation, as when sexual harassment is allowed to occur.
4. The informal norms that develop within a bureaucracy may cause worker productivity to change, dependent upon the norms and how they are informally enforced. In the Hawthorne Studies, for example, researchers found that workers who produced too much or too little each day were labeled or ridiculed by others.

D. Problems of Bureaucracies
   1. Several problems, including the risky shift in work groups and the development of groupthink, develop from the nature of the complex bureaucracy.
   2. **Ritualism**. Another problem in bureaucracies is *organizational ritualism*, which refers to workers' rigid adherence to the rules, regardless of whether their behavior accomplishes the purpose for which the rule was originally designed.
      a. The 1986 explosion of the Space Shuttle *Challenger* and the 2003 break-up of the Space Shuttle *Columbia* are both examples of ritualism or what was called a "flawed institutional culture" and a normalization of deviance accompanying a gradual erosion of safety margins.
      b. Such rigid group conformity within an organizational setting can lead to deviant behavior being redefined so that it is perceived as normal.
   3. **Alienation**. Alienation also develops within bureaucracies. It is characterized by the individual becoming psychologically separated from the organization and its goals, which may result in increased turnover, tardiness, absenteeism, and overall dissatisfaction with the organization.

E. The McDonaldization of Society
   1. George Ritzer used the term the *McDonaldization of society* to describe the phenomenon whereby the principles that characterize fast food organizations increasingly dominate more aspects of American society. Ritzer's theory is based on Weber's prediction that human behavior would be increasingly guided by rational systems—rules, regulations, and formal structures—rather than by abstract values. The four features of McDonaldization are:
      a. **efficiency**: things move from start to completion in a standardized, streamlined way, often with machines or customers doing work once done by an employee;
      b. **calculability**: emphasis is placed on the quantitative aspects of products sold, including size, cost, and time of production, rather than quality;
      c. **predictability**: assurance that products will be exactly the same, regardless of where or when purchased, is offered; and
      d. **control**: the primary organizational principle of McDonaldization, control refers to the reduction of people's behavior to a series of machine-like actions. Both customers and workers are carefully monitored in these organizations.

2. McDonaldization brings benefits, such as greater availability of goods and services to a wider portion of the population and instantaneous service and convenience to a public with less free time. Additionally, there is predictability and familiarity in the goods bought and sold, and standardization of pricing and uniform quality of the goods.

3. McDonaldization also has disadvantages, such as a danger of dehumanization of workers and the threat of loss of creativity as people accept increasingly standardized products and services.

V. DIVERSITY: RACE, GENDER, AND CLASS IN ORGANIZATIONS

Since organizations tend to reflect patterns within the broader society, the hierarchical structure of positions is marked by inequality in race, gender, and class relations. Women and minorities confront a "glass ceiling" effect. The more egalitarian the firm's environment, the more equitable will be its treatment of women and minorities. Rosabeth Moss Kanter's research shows how the structure of organizations leads to obstacles in the advancement of groups that are tokens in the organizational environment. Such groups experience great stress and have difficulties gaining credibility not only with their superiors, but with their co-workers as well. Social class also plays a part in determining people's place within formal organizations. In response to a changing workforce, diversity training has become commonplace in most large organizations.

VI. FUNCTIONAL, CONFLICT, AND SYMBOLIC INTERACTION: THEORETICAL PERSPECTIVES

The functionalist perspective focuses on the positive functions—called *eufunctions*—of bureaucracies, as well as the problems of bureaucracies—their dysfunctions. The conflict perspective argues that the hierarchical nature of bureaucracy encourages conflicts between superior and subordinate, as well as conflicts based on race, ethnicity, gender, and social class. The symbolic interaction perspective underlies the theories of Argyris (which focuses on self-actualization as a way of reducing organizational dysfunctions) and Ouchi (which argues for increased supervisor-subordinate interaction).

INTERNET EXERCISES

1. Have students search the Internet for McDonaldization. They will encounter several home pages, some produced for sociology classes and others with a more activist bent. Some of the sites that seem the most interesting include (www.mcdonaldization.com/), which has become a blog; (www.geocities.com/mcdonaldization/), created by Mariya Levin, a student who observes McDonaldization in society and in her life; and McSpotlight (www.mcspotlight.org/), formed in 1996 around the McLibel trial in England and operated by an independent network of volunteers working in 16 countries.

2. Visit the website for Robert Putnam's book, *Bowling Alone*, at (BowlingAlone.com) to access research and community action ideas. The site includes the data set for the book and an extensive bibliography, as well as selections and facts presented in the book itself. There is a companion site that can be reached through the home page on the Better Together report (bettertogether.org), which allows you to join a ListServ and to explore ways of building community and extending civic engagement in your own community. Report on one new insight or idea for extending civic engagement you learned from your exploration.

INFOTRAC EXERCISES

Conduct a keyword search using InfoTrac College Edition to extend the discussion in DOING SOCIOLOGICAL RESEARCH: *Sharing the Journey* (p. 128-129).

1. **Keyword: Small group interaction.** This search leads to articles on small-group learning, peer interaction, race relations, and reciprocity.

2. **Keyword: Support group.** This search leads to consideration of support groups related to gambling, the unemployed, minority student retention, medical conditions, and life-style change.

3. **Keyword: Civic participation or engagement.** This search finds articles on civic life, education, empowerment, and discussions and debates on Putnam's thesis in Bowling Alone.

4. **Keyword: Voluntary organizations.** Search results deal with civil society, dynamics within voluntary organizations, and the relation of voluntary organizations to government.

PRACTICE TEST

Multiple Choice Questions

1. Which of the following statements about juries is (are) **true**?
   a. As groups, juries behave differently than do the individual members who comprise each jury.
   b. Sociologists can make educated predictions about who will become the most influential member in a jury.
   c. Jury verdicts correlate not only with the evidence, but also with the racial-ethnic and gender composition of the jury.
   d. all of these choices

2. A group is a collection of individuals who _____.
   a. regularly interact with each other
   b. share a subjective sense of "we"
   c. share norms and goals
   d. all of these choices

3. Simmel identified the tendency for triads to separate into a pair and an isolate as triadic _____.
   a. aggregation
   b. segregation
   c. alignment
   d. isolation

4. A group consisting of intimate, face-to-face interaction and relatively long-lasting relationships, such as a family, is a(n) _____.
   a. secondary group
   b. social category
   c. primary group
   d. audience

5. Which of the following statement about secondary groups is true?
   a. They serve primarily instrumental or task-oriented needs.
   b. They generally have fewer members than do primary groups.
   c. They tend to have long-lasting, powerful influences on their members' self-concepts.
   d. none of these choices

6. Although James has never met or interacted with a professional athlete, he reveres major league baseball players and uses them as a standard for determining his own values about physical fitness and good sportsmanship. For James, professional athletes represent a _____ group.
   a. secondary
   b. reference
   c. primary
   d. ideal

7. Because of attribution error, when people observe an *in-group* member behaving in an exceptional manner, they attribute the behavior to the _____.
   a. person's good luck
   b. influence of a supreme deity
   c. situation in which the behavior occurs
   d. person's good disposition or "true nature"

8. When a person is prejudiced toward a particular *out-group* and observes a member of that group behaving in a positive, non-stereotypical manner, the observer will probably _____.
   a. dismiss his/her prejudice about the out-group because s/he realizes that the stereotype is based on inaccurate information
   b. retain his/her original prejudice because s/he will assume that the behavior was a special exception to the person's usual behavior
   c. attempt to become a member of the out-group so that s/he can better understand the members' perspectives and experiences
   d. consult members of his/her in-group about the observation to evaluate the new information more objectively

9. The set of links between individuals or other social units, such as bureaucratic organizations or entire nations, is referred to as a _____.
   a. global form
   b. social network
   c. social organization
   d. dispositional network

10. The tendency for group members to reach a consensus opinion, even if the decision is stupid, is referred to as _____.
    a. generalization
    b. polarization
    c. groupthink
    d. attribution

11. Wuthnow's study of small group membership in the United States indicates that _____.
    a. membership in formal religious organizations has increased as the society has become more remote and alienating
    b. very few people voluntarily participate in any type of small group today because these groups threaten American's strong sense of individuality and personal freedom
    c. a growing number of people have joined recovery and reading groups in their communities to garner emotional support in an increasingly impersonal society
    d. the common assumption that geographic mobility and modernization have eroded the family's role in providing a sense of belonging and integration for individuals is a myth

12. The Marriot Corporation now provides the same food to students and staff at over 500 college campuses across the United States. George Ritzer suggests that the predictability and uniformity associated with these meals is due to the process known as the _____.
    a. bureaucratization of society
    b. McDonaldization of society
    c. monopolization of society
    d. ritualization of society

13. According to Goffman, which of the following statements about people who enter total institutions is (are) **true**?
    a. Most people who enter total institutions do so voluntarily.
    b. The main reason that people enter total institutions is to gain personal satisfaction.
    c. When people enter total institutions, they are subjected to practices designed to make them surrender their former identities.
    d. all of these choices

14. According to Max Weber's ideal type, which of the following is (are) typical of a bureaucracy?
    a. Extensive, detailed lists of rules and procedures
    b. Cooperative, non-hierarchical management styles
    c. Low degree of division of labor and specialization
    d. all of these choices

15. The informally evolved culture that develops over time as a reaction to the formality and impersonality of bureaucracy represents _____.
    a. the McDonaldization of society
    b. bureaucracy's other face
    c. the small world problem
    d. groupthink

16. The explosion of the Space Shuttle *Challenger* and the break-up of the Space Shuttle Columbia both seemed to be the result of an organizational problem known as _____.
    a. ritualism
    b. alienation
    c. rate busting
    d. rationalization

17. Which common problem is developing when individuals become psychologically separated from a bureaucratic organization and its goals and there is an increase in tardiness and absenteeism?
    a. ritualism
    b. alienation
    c. rate busting
    d. rationalization

18. According to George Ritzer, the McDonaldization of society is characterized by an increasing emphasis on the _____.
    a. level of creativity encouraged by management
    b. range of choice provided to the customers
    c. quantity of the products created
    d. quality of the products created

19. The torture and abuse of Iraqi prisoners in Abu Ghraib prison by U.S. military personnel demonstrates that _____.
    a. American soldiers lack education and are thus susceptible to going along with improper and unethical behavior
    b. the solders who engaged in the torture were a few "rotten apples" who had sadistic personalities
    c. the causes of the soldiers' behavior lie in the social structure and group pressures of the situation
    d. the soldiers were influenced by Iraqi culture and so abandoned their Western values

20. Studies on race and gender within formal organizations find that _____.
    a. the hierarchical structuring of positions within organizations discourages gender and racial discrimination
    b. women and racial minorities are treated equally in formal organizations, based upon their level of education and professionalism
    c. patterns of race and gender discrimination persist throughout organizations even when formal barriers to advancement have been removed
    d. once women and minorities are hired, they experience equal treatment with respect to job retention and promotion

21. Which of the following characteristics increases the likelihood that groupthink will occur?
    a. a full discussion of all possible options
    b. an illusion of invulnerability
    c. an illusion of disunity
    d. all of these choices

22. The Black sorority known as the Deltas is a(n) _____ organization whose members voluntary join the organization primarily for personal fulfillment.
    a. total
    b. coercive
    c. utilitarian
    d. normative

True-False Questions

1. The size of a group has relatively little effect on how members of the group behave.

   TRUE or FALSE

2. In times of crisis or high stress, such as a natural disaster, some secondary groups can take on the characteristics of primary groups.

   TRUE or FALSE

3. Simmel noted that a triad is inherently unstable, whereas dyads are relatively stable groups.

   TRUE or FALSE

4. In his *risky shift* experiments, Stoner found that Americans are less likely to take greater risks as members of a group than as individuals.

   TRUE or FALSE

5. A total institution is a charitable organization such as NOW or the NAACP, that people voluntary join because they support the organizations' goals.

   TRUE or FALSE

6. In his conformity studies, Milgram found that the race and sex of the subjects had no significant effect on whether they would comply with the request to administer shocks during the experiment.

   TRUE or FALSE

7. Choices of political party and religious affiliation in adulthood correlate strongly with those of one's parents.

   TRUE or FALSE

8. Informal norms within a bureaucracy are almost always beneficial to management because they usually contribute to increased productivity.

   TRUE or FALSE

9. Research indicates that people get high-paying, prestigious jobs through formal job listings, want ads, and placement agencies more often than through personal networks.

   TRUE or FALSE

10. Research shows that Black federal employees are more than twice as likely to be dismissed as their White counterparts primarily because they have lower educational levels and don't perform as well in their jobs.

TRUE or FALSE

Fill-in-the-Blank Questions

1. Risky shift is caused partly by _____ , a process in which a person perceives his or her own identity as merging with the group.

2. A _____ group is a group to which you may or may not belong, but that you use as a standard for evaluating your values, attitudes, and behaviors.

3. A _____ is an organization cut off from the rest of society in which individuals who reside there are subject to strict social control.

4. Social interactions in bureaucratic settings that ignore, change, or otherwise bypass the formal structure and rules of the organization are called _____ .

Essay Questions

1. Analyze the McDonaldization of some sector of U.S. society, like the healthcare system, shopping malls, your college or university.

2. Using an example, explain how risky shift occurs.

3. If given the opportunity, what advice would you offer the Presidential Cabinet to prevent groupthink?

4. Identify the six characteristics of bureaucracy according to Weber and explain how each feature supports the goals of bureaucratic organizations.

5. Imagine that you are the CEO of a telemarketing organization. What steps would you take to counter worker alienation and ritualism?

Solutions

PRACTICE TEST

Multiple Choice Questions

1. D, 125-6 Juries are groups, and groups behave differently than individuals. Predictions can be made about who will become the most influential jury member, and jury composition is related to the verdict rendered.

2. D, 126 A group is a collection of individuals who interact with each other and share goals and norms. A group also has a subjective awareness of "we." By contrast, an audience is an example of a social category, rather than a group.

3. B, 127 A triad is an unstable group with a tendency to segregate into a pair (a dyad) and an isolate, resulting in triadic segregation.

4. C, 127 According to Cooley, a group consisting of intimate, face-to-face interaction and relatively long-lasting relationships is a primary group. It usually consists of fewer members than a secondary group, and is concerned primarily with meeting members' expressive needs.

5. A, 128 Secondary groups are generally less permanent and larger than primary groups and primarily serve members' instrumental needs.

6. B, 129 Reference groups are those to which you may or may not belong, but that you use as a standard for evaluating your values, attitudes, and behaviors.

They may have both positive and negative influences on individual behavior.

7. D, 130 Attribution errors include the following: If an in-group member performs in an exceptional way, they are given credit for a valued disposition. Also, when observing improper behavior by an out-group member, onlookers are likely to attribute the deviance to the disposition of the wrong doer. If an out-group member performs in an exceptional way, it is often attributed to good luck or the individual is viewed as "exceptional." When observing improper behavior by an in-group member, onlookers are likely to attribute the behavior to the situation of the wrong doer.

8. B, 130-1 If an out-group member behaves in an exceptional way, the person is often perceived as an exception to the rule or as exceptionally lucky.

9. B, 131 A set of links between individuals or between other social units, such as bureaucratic organizations or entire nations, is a social network. Social network membership varies by gender, race, and class.

10. C, 135 Janis argues that groupthink results from an illusion of invulnerability, a falsely negative impression of those who are antagonistic toward the group's plans,

discouragement of dissenting opinion, and an illusion of unanimity.

11. C, 128-9 Wuthnow's research on small group participation indicates that an increasing number of people have joined small recovery, reading, and spiritual groups in their search for emotional support, commitment, and meaning. These groups provide a sense of belonging and integration in an otherwise impersonal society, and individuals are free to leave the group if it no longer meets their needs. The role of the family as a source of integration has been eroded by several fundamental changes in society, including geographic mobility.

12. B, 141-2 George Ritzer identified four dimensions of the McDonalidization process: efficiency, calculability, predictability, and control.

13. C, 138 Goffman referred to coercive organizations as total institutions, such as prisons and mental hospitals. Most people enter these organizations involuntarily and must surrender their former identities and rights to privacy, which is facilitated by subjecting them to "degradation ceremonies."

14. A, 138-9 A bureaucracy is an organizational form characterized by an authority hierarchy, clear division of labor, explicit rules, and impersonality.

15. B, 139-40 The informally evolved culture that develops over time as a reaction to the formality and impersonality of the bureaucracy is bureaucracy's other face. McDonaldization refers to the tendency for features of the fast-food business to be adopted in the larger society. The small world problem refers to the high density of social networks in the U.S., particularly among African Americans. Groupthink refers to the tendency of groups to urge consensus among the participants, even if a poor decision is the result.

16. A, 140 Problems that may develop from the nature of bureaucracy include alienation, ritualism, and groupthink. The Challenger and Columbia accidents reflect workers' and managers' rigid adherence to rules, or organizational ritualism.

17. B, 140-1 Alienation may be widespread in organizations where individuals have little control over their work and engage in repetitive tasks. It results from worker's psychological separation from the organization and its goals.

18. C, 141-2 McDonaldization stresses quantity over quality of products. It involves greater control, resulting in less worker autonomy and fewer choices for customers.

19. C, 134-5 The torture in Abu Ghraib was predicted by research on social influence and obedience, to lie not in the personalities or educational level of the soldiers, but rather in the social structure and group pressures of the situation.

20. C, 142-3 Studies on race and gender within formal organizations find that patterns of race and gender discrimination persist throughout the organization even when formal barriers to advancement have been removed. The hierarchical structuring of such organizations results in the concentration of power and influences with a few individuals, with Whites and males more likely to be promoted and promoted more quickly. Women and minorities tend to confront a "glass ceiling" to promotion, and Black federal employees are more than twice as likely to be dismissed as their White counterparts.

21. B, 135 Groupthink, the tendency for group members to reach a consensus opinion, is likely to occur when there is an illusion of invulnerability, a falsely negative impression of those who are antagonists to the group's plans, discouragement of dissenting opinion, and an illusion of unanimity.

22. D, 137-8 People generally join normative, or voluntary, organizations to pursue personal fulfillment, and utilitarian organizations for rewards such as wages. Membership in coercive organizations is usually involuntary.

True-False Questions

1. F, 127 Simmel's studies of dyads and triads demonstrated that group size influences group behavior.

2. T, 128 Secondary groups may take on the characteristics of primary groups in certain social situations.

3. T, 127 Simmel noted that a triad is an unstable grouping that tends to separate into a dyad (pair) and an isolate (individual).

4. F, 136 Americans are more likely to take risks, such as streaking, when they are in groups because they experience deindividuation, or the merging of the self with the group, coupled with a decreased sense of personal responsibility.

5. F, 138 Goffman referred to coercive organizations, such as prisons and mental hospitals, as total institutions. People do not usually join these organizations voluntarily. People voluntarily join charitable, or normative, organizations, such as NOW and the NAACP, for personal fulfillment.

6. T, 133-4 Milgram did not detect any significant effects of race, class, or gender on the research subjects' behavior in his studies of conformity.

7. T, 127 Primary groups such as the family have a powerful influence on people's values and behavior throughout their lives.

8. F, 139-40 The informal norms that develop as a result of bureaucracy's other face may lead to exclusionary practices and reduced worker productivity.

9. F, 131 Research indicates that people get jobs, especially high-paying, prestigious jobs, via personal networks rather than through formal job listings.

10. F, 143 Studies show that Black federal employees (men and women) are more than twice as likely to be dismissed as their White counterparts, regardless of education, occupational category, pay level, type of federal agency, age, performance rating, seniority, or attendance record. The main reasons cited in the studies for these disparities were lack of network contacts and racial bias within the organization.

Fill-in-the-Blank Questions

1. deindividuation, 137
2. reference, 129
3. total institution, 138
4. bureaucracy's other face, 139-40

Essay Questions

1. McDonaldization. The characteristics of McDonaldization are discussed on p. 141-142. Student examples should describe the societal sector in terms of the four characteristics of McDonaldization: efficiency, calculability, predictability, and control.

2. Risky Shift. The dynamics of risky shift are described on p. 136-137. Student examples should illustrate these dynamics.

3. Groupthink. The several characteristics that outbreaks of groupthink share are described in the text on p. 135-136. Student advice to the Presidential cabinet should address all four characteristics.

4. Bureaucracies. The six characteristics of bureaucracy are discussed on p. 139 of the text. Student responses should explain how each of these characteristics supports the goals of bureaucratic organizations.

5. Alienation and Ritualism. The conditions in bureaucracies that give rise to alienation and ritualism are described on p. 140-141. Student responses should take these situations into account and propose ways of reducing alienation and ritualism.

CHAPTER 6

DEVIANCE AND CRIME

BRIEF CHAPTER OUTLINE

CHAPTER FOCUS

This chapter examines deviance from a sociological perspective, including the contributions of major theories to the understanding of deviance. The chapter also explores the types of crime, patterns of treatment within the criminal justice system in terms of race, gender, and class, and major trends and problems within the criminal justice system. Globalization of crime, especially terrorism, is also addressed.

QUESTIONS TO GUIDE YOUR READING

1. What are the differences between sociological and psychological perspectives on deviance?

2. What are the strengths and weaknesses of functionalist theories of deviance, including Durkheim's approach to the study of suicide, Merton's structural strain theory, and social control theory?

3. How do conflict theories approach deviance, elite deviance, and social control?

4. How do symbolic interaction theories, including W. I. Thomas and the Chicago School, differential association and labeling theory, help us to understand deviant careers and communities and to critique problems with official statistics?

5. What is the correlation between social status and mental illness, and how do social stigmas develop into master statuses?

6. How do societal perceptions and reactions vary with respect to the major types of crimes—personal and property, hate, victimless, corporate, and organized?

7. How is the treatment people experience within the criminal justice system (police, courts, and the law) significantly related to patterns of race, gender, and class inequality?

8. What explains the high incarceration rate in the US, and what is the impact of the privatization of prisons?

9. Do prisons rehabilitate or simply warehouse prisoners?

10. How is crime—particularly terrorism—globalized?

KEY TERMS

(defined at page number shown and in glossary)

altruistic suicide 154

anomie

crime 164

cyberterrorism 177

deviant career 160

deviant identity 160

egoistic suicide 155

hate crime 167

labeling theory 159

medicalization of deviance 153

personal crimes 167

racial profiling 173

social control agents 158

stigma 163

KEY PEOPLE

(identified at page number shown)

CHAPTER OUTLINE

I. **DEFINING DEVIANCE**

   A. <u>Sociological Perspectives on Deviance.</u> The sociological definition of **deviance**—as *behavior that is recognized as violating expected rules and norms—stresses social context, recognizes that not all behaviors are judged similarly by all groups*, and *recognizes that established rules and norms are socially created.* Sociologists distinguish between *formal* (behavior that breaks laws or official rules) and *informal deviance* (behavior that violates customary norms). Sociologists study both why people violate norms and how society reacts. *Labeling theory* focuses on the role of society in *creating* deviance.

   1. **The Context of Deviance.** The variation in definitions of deviance and punishment of deviance by context and overtime points to the social function of deviance for the society. Durkheim argued that societies need deviance to know what presumably normal behavior is. Recognition and punishment of deviance affirms the collective beliefs of the society, reinforces social order, and inhibits future deviant behavior. Deviance thereby produces social solidarity.

   2. **The Influence of Social Movements.** Social movements, organized by moral entrepreneurs, can reform how a behavior is perceived and dealt with by society (e.g., the anti-smoking movement and the gay and lesbian movement).

   3. **The Social Construction of Deviance.** The public commonly understands deviance as the result of individualistic or personality factors. Deviance may be a positive adaptation to a situation (the Andes survivors) or as a rational adaptation to some situations (gang membership). Often deviant behavior is encouraged and praised, and some behaviors defined

as deviant are similar to so-called normal behavior. What is considered deviant or normal depends on the social context.

B. Psychological Explanations of Deviance
1. Sociologists have been critical of psychological explanations of deviance, which emphasize individual factors as the underlying cause of deviant behavior, because they overlook the context in which deviance is produced.
2. The **medicalization of deviance** refers to the popularity of explanations of deviant behavior that interpret deviance as the result of individual pathology, an approach that focuses on the physical or genetic roots of deviant behavior.
3. Sociologists integrate individual factors into explanations of deviance that emphasize the social context of the behavior. For example, although alcoholism has medical consequences, successful treatment requires that the social relationships, social conditions, and social habits of alcoholics be altered.

II. **SOCIOLOGICAL THEORIES OF DEVIANCE**

A. Functionalist Theories of Deviance

Functionalist theory interprets all parts of society, including deviance, as contributing to the stability of the whole. Deviance creates social cohesion, clarifies society's norms, and affirms the collective identity of the group when those who are defined as deviant are labeled or condemned.

1. *Durkheim: The Study of Suicide.* **Emile Durkheim** developed his analysis of deviance through his study of suicide, in which he was critical of psychological interpretations of suicide. He developed an alternative, sociological explanation that emphasized the role of social structure in producing deviance, noting that suicide is a social phenomenon.
   a. Durkheim investigated causes of suicide related to time and place, rather than emotional stress, and identified the importance of social attachments in preventing or producing deviant behavior such as suicide.
   b. Durkheim analyzed three types of suicide: anomic suicide, altruistic suicide, and egoistic suicide.
      i. **Anomie** is the condition which exists when social regulations in a society breakdown so that people exist in a state of relative normlessness. **Anomic suicide** occurs when the disintegrating forces in society make individuals feel lost or all alone, as in the case of students on college campuses who commit suicide.
      ii. **Altruistic suicide** occurs when there is excessive regulation of individuals by social forces. For example, the terrorists who hijacked and crashed 4 airplanes during the September 11 tragedy were willing to kill themselves to achieve their goals.
      iii. **Egoistic suicide** occurs when people feel totally detached from society. For example, the high rate of suicide among elderly men in the United States, who often lose their work roles and have weakened ties to family and community, may be egoistic.
   c. The concept of anomie has implications beyond suicide. Although anomie is reflected in how individuals feel, its origins are in society, where there may be unclear or conflicting norms. Anomie is related to a variety of social problems, including deviant behavior.

2. ***Merton: Structural Strain Theory.*** **Robert Merton** (1910-2003) further developed the functionalist perspective on deviance by his **structural strain theory** traces the origins of deviance to the tensions caused by the gap between cultural goals and the means people have to achieve those goals.

   a. Societies are characterized by *culture*, which establishes goals for both people and society, and *social structure*, which provides (or fails to provide) the means for people to achieve these goals.

   b. When the means are out of balance with the goals, deviance is likely to occur. For example, prostitutes accept the goals of society but lack the means to achieve those goals, so they develop creative, although illegitimate, means of economic support.

3. ***Social Control Theory.*** **Social control theory**, developed by Travis Hirschi, posits that deviance occurs when an individual's or group's attachment to social bonds is weakened. This theory, which assumes there is a common value system within society and that deviance derives from breaking one's allegiance to that system, can help explain why 2 teen boys committed murder and suicide at a Littleton, Colorado school.

4. ***Functionalism: Strengths and Weaknesses.*** Functionalists emphasize that social structure, not just individual motivation, produces deviance, and deviance may be functional for society. Although individuals choose to behave in a deviant manner, they make their choices from among socially structured options. Critics of functionalism argue that it does not explain how norms of deviance are established, nor does it explain why some behaviors are defined as normative and others as illegitimate.

B. Conflict Theories of Deviance

   1. Another *macrostructural* approach, conflict theory links the study of deviance to power relationships and social inequality. Crime committed among the poorest groups is viewed as the result of the economic status of those groups, and the ability of those who engage in **elite deviance** to hide their crimes (such as *corporate crime* or *white-collar crime*) and avoid criminal labels as the result of their access to political and economic power. Conflict theorists also argue that the law is created by elites to protect the interests of the dominant class and to regulate populations that pose a potential threat to affluent interests, and to bring them into conformity through **social control. Social control agents**, such as police and mental health workers, have the power to label less powerful people, as evident in the patterns of arrest data.

   2. ***Conflict Theory: Strengths and Weaknesses.*** Conflict theory provides insight into the significance of power relationships in the definition, identification, and handling of deviance, and offers a powerful analysis of how injustices produce crime and result in different systems of justice for disadvantaged and privileged groups. On the other hand, conflict theory is less effective in explaining forms of deviance other than crime, or forms of deviance not involving economic interests.

C. Symbolic Interaction Theories of Deviance

   1. ***W. I. Thomas and the Chicago School.*** Symbolic interaction theories are microstructural theories that emphasize how deviance originates in the interaction between different groups and is defined by society's reaction to certain behaviors. **W. I. Thomas** (1863-1947) developed

situational analysis, a way to explain deviances as *a normal response to the social conditions in which people find themselves.*

2. ***Differential Association Theory*** emphasizes the significant role that peers play in encouraging deviant behavior and explains why deviant activity may be more common in some groups than others.

3. ***Labeling Theory*** linked with conflict theory shows how those with the power to label an act or a person as deviant and impose sanctions wield great power in determining societal understandings of deviance. Labeling theory distinguishes between *primary*, *secondary*, and *tertiary deviance*. The formation of a **deviant identity** involves a process of social transformation in which a news self-image and new public definition of a person emerges.

4. ***Deviant Careers*** are studied like any other careers, including the progression through deviance, how they are sustained through people's reactions, and how deviants experience career mobility.

5. ***Deviant Communities***, like subcultures and countercultures, maintain their own values, norms, and rewards for deviant behavior, and some communities provide support to those in presumed deviant categories, like various "12 step" programs.

6. ***A Problem with Official Statistics.*** Reported rates of deviant behavior are themselves the product of the socially determined behavior of identifying what is deviant, a product influenced by power relationships and social inequalities.

7. ***Labeling Theory: Strengths and Weaknesses.*** Labeling theory illuminates the consequences of deviance, but it does not explain the origins of deviant behavior. The failure of labeling theory to explain official crime and deviance statistics is corrected by conflict theorists by placing their analysis of deviance with the power relationships of race, class, and gender.

III. FORMS OF DEVIANCE

A. <u>Mental Illness</u>

Sociologists give two explanations for the correlation between social status and mental illness—the harsher social environment experienced by people with low-incomes, racial minorities, and women in a sexist society, and the greater likelihood that people with the fewest resources will be labeled mentally ill. Both explanations are substantiated in studies of the homeless, racial minorities, and women

B. <u>Social Stigmas</u>. A **stigma**, like a disability, can become a **master status**, overriding all other features of the person's identity. As Goffman put it, people with stigmas are perceived as having a spoiled identity. Stigmatized individuals may try to hide their stigma or blame others.

C. <u>Substance Abuse: Drugs and Alcohol</u>. Social factors influence drug and alcohol use and definitions of deviance, as illustrated by changing attitudes toward cigarette smoking. Patterns of drug and alcohol use vary by factors such as age, gender, and race.

IV. CRIME AND CRIMINAL JUSTICE

A. <u>Measuring Crime: How Much Is There?</u> **Criminology** is the study of **crime**, or *deviance* that violates specific criminal laws.

1. ***Personal and Property Crimes.*** The FBI's *Uniform Crime Reports* includes **personal crimes** (murder, aggravated assault, forcible rape, and robbery) and **property crimes** (burglary, larceny, auto theft, and arson). The rate of violence crime has actually been declining somewhat

in recent years. **Hate crimes** have increased in recent years. **Victimless crimes**, not listed in the FBI's serious crime index, include activities such as gambling, illegal drug use, and prostitution, in which there is no complainant.

2. ***Elite and White-Collar Crime.*** Although *white-collar crime* does not generate public interest, and is generally the least investigated and least prosecuted, in terms of costs it is even more consequential for society. White-collar criminals tend to "normalize" their behavior.

B. <u>Organized Crime</u>. **Organized crime** syndicates, typically involved in the provision of illegal goods and service to others, are often based on racial and ethnic membership and are organized along the same lines as legitimate businesses.

C. <u>Corporate Crime and Deviance: Doing Well, Doing Time</u>. Sociologists estimate that the costs of corporate crime may be as high as $200 billion every year, compared with roughly $15 billion from street crime. Tax cheaters alone skim an estimated $50 billion annually from the IRS. *Corporate crime and deviance* is institutionalized deviance within the context of a formal organization or bureaucracy that is actually sanctioned by the norms and operating principles of the bureaucracy, through a process known as the *normalization of deviance*.

D. <u>Race, Class, Gender, and Crime</u>. Socioeconomic status correlates with crime, but prosecution by and treatment within the criminal justice system—from official arrest statistics, treatment by policy, patterns of sentencing, to imprisonment—is significantly related to patterns of race, gender, and class inequality.

1. ***Race, Class, and Crime.*** The poor are more likely than others to be arrested for crimes. There is a strong correlation between unemployment, poverty, and crime, but law enforcement is also concentrated in lower-income and minority areas. Police concentrate on street crime, not white-collar crime. The poor are more likely to be prosecuted, convicted, and sentenced to prison. For examples, in the juvenile court system, 35% of arrested White youths are sent to prison, but nearly 60% of Black youths are imprisoned. In addition to the factors that affect official rates of arrest and conviction—bias of official statistics, influence of powerful individuals, discrimination in patterns of arrest, differential policing—there remains evidence that actual commission of crimes varies by race. Racial minority groups are far more likely than Whites to be poor, unemployed, and living in single-parent families, which are all predictors of a higher rate of crime.

2. ***Gender and Crime.*** Although women's participation in crime has increased in recent years, their crime and arrest rates are still small relative to men, except for crimes such as fraud, embezzlement, and prostitution. Women are also less likely than men to be victimized by crimes, although they are more fearful of crime than men. Victimization by rape—the fastest growing and most underreported crime—is probably the greatest fear. The most powerless women—the young, minority, poor, divorced, single, or separated women—have the highest likelihood of being raped.

E. <u>The Criminal Justice System: Police, Courts, and the Law</u>.

1. ***The Policing of Minorities.*** Minority communities are policed more heavily than White neighborhoods, and policing in minority communities has a different effect than in White middle-class communities.

a. Police, using **racial profiling**, are more likely to stop, search, and use force against minority drivers; and racial minorities are more likely than the rest of the population to be victims of excessive use of force by the police.

b. Those cities where African Americans head the police department show a concurrent decline in police brutality complaints, an increase in minority police and minority recruitment, and in some cases, a decrease in crime.

c. Increasing the number of African Americans in police departments has some positive effect but does not change the material conditions that create crime.

2. *Race and Sentencing.* African Americans and Latinos experience higher bails, have less success in plea bargaining, are found guilty more often, are likely to get longer sentences for the same crimes, are less likely to be released on probation, and receive longer sentences than White Americans. Sentencing also differs depending on the racial identity of the judge. Racial discrimination is particularly evident with respect to the death penalty. Forty-two percent of about 3,500 prisoners on Death Row are Black, which is largely the result of discriminatory sentencing.

3. *Prisons: Rehabilitation or Warehousing?* The criminal justice system reflects the racial and class stratification in society.

a. Racial minorities are more than half of the federal and state male prisoners in the U.S. The U.S. and Russia have the highest rate of incarceration in the world and the rate of imprisonment in the U.S. has been rapidly growing, at an annual cost of at least $150 billion.

b. This high rate of growth has led to increased privatization in the operation of prisons, a trend that raises many social policy questions, including the fact that *the rate of violence in private prisons is higher than in state facilities*. Major reasons for the growth in the prison population are increased enforcement of drug offenses and mandatory sentencing.

c. The rate of imprisonment of women is also increasing, a trend that poses many problems in a system designed for men and run mostly by men. About *two-thirds* of all women prisoners have been victims of sexual abuse.

d. Research indicates that prisons neither deter nor rehabilitate offenders. Some radical sociologists contend that the criminal justice system is not meant to reduce crime, but rather to reinforce an image of crime as a threat from the poor and from racial groups.

V. **TERRORISM AS INTERNATIONAL CRIME: A GLOBAL PERSPECTIVE**

Terrorism, defined by the FBI has the unlawful use of force or violence against persons or property to intimidate or coerce a government or population in furtherance of political or social objectives, is one example of the globalization of crime. **Bioterrorism** and **cyberterrorism** (www.cybercrime.gov/) are specific forms of terrorism. Sociologists look to the social structure of conflicts from which terrorism emerges as the cause of such crime and deviant behavior.

INTERNET EXERCISES

1. The Federal Bureau of Investigation website (www.fbi.gov/) is particularly useful for its crime statistics and reports information. This includes the Uniform Crime Reports, as well as reports on hate crimes,

terrorism, family violence, corporate fraud, and initiatives like SafeStreets. Conduct research on one crime of interest to you, and right a brief analytic report on the dimensions and trends in this crime.

2. The Bureau of Justice Statistics website (www.ojp.usdoj.gov/bjs/) provides interesting data, reports, and analyses on a number of topics, including crime victimization, criminal offenders, courts and sentencing, firearms and crime, homicide rates, reentry trends, and international crime statistics. Use the site as assigned for class reports. Note the usefulness of comparing crime victimization data with crimes reported in the UCR.

3. The Sentencing Project (www.sentencingproject.org/) conducts extensive studies on the criminal justice system, with particular attention to sentencing and other policies related to the poor, minorities, and women, as well as felony disenfranchisement and community advocacy. You may find this site useful in class assignments or research projects on particular topics.

INFOTRAC EXERCISES

Conduct a keyword search using InfoTrac College Edition to extend the discussion in TAKING ON SOCIAL ISSUES: *Should Drugs Be Decriminalized?* (p. 164)

1. Keyword: Decriminalization of drugs. This search yields articles on legalization of illicit drugs, the effects of decriminalization, and so on.

2. Keyword: Drug policy. This search yields articles on drug testing, drugs in sports, anti-drug media efforts, and priorities.

3. Keyword: War on drugs. In this search, articles are found contrasting and linking the war on drugs and the war on terror, medical marijuana, war and democracy, street drugs, impact on source countries, casualties, and so on.

PRACTICE TEST

Multiple Choice Questions

1. Sociologists criticize the medicalization of deviance on which of the following grounds?
    a. Viewing solely the medical aspects of deviant behavior ignores the effects of social structure on the development of the behavior.
    b. Deviance is not a pathological state, but an adaptation to the social structure within which people live.
    c. Medicalized views of deviance focus on curing the "sick" state of mind, rather than changing the social conditions that gave rise to the deviant behavior.
    d. all of these choices

2. Body piercing violates customary norms, so it is an example of which type of deviance?
   a. elite
   b. formal
   c. medical
   d. informal

3. According to conflict theory, crime results primarily from the _____.
   a. competition for social resources in capitalist societies
   b. lack of shared norms and social integration in modern societies
   c. individual genetic deficiencies that result in psychological impairment
   d. cultural transmission of deviant values from one generation to the next

4. According to the FBI, assaults and other malicious acts motivated by various forms of bias are _____.
   a. organizational deviance
   b. victimless crimes
   c. elite deviance
   d. hate crimes

5. Criminologist Joel Best argues that crimes, such as "wilding" and "road rage" _____.
   a. are completely random phenomena
   b. have markedly increased since 1990
   c. are patterned, predictable social random acts of violence
   d. are unrelated to the social characteristics of the victims and the perpetrators

6. The condition that exists when social regulations break down in a society and people exist in a state of relative normlessness is _____.
   a. crime
   b. anomie
   c. altruism
   d. deviance

7. According to Durkheim, which type of suicide occurs when people feel totally detached from society due to weakened social ties?
   a. anomic
   b. egoistic
   c. altruistic
   d. egalitarian

8. Which of the following statements about crime in the U.S. is **true**?
   a. The crime rate has increased steadily throughout the 1990s, as part of a decades-long pattern leading to unprecedented levels of crimes today.
   b. Official crime statistics tend to give a biased picture of crime, by focusing on violent and property crimes and under-representing elite crimes.
   c. National Crime Victimization Surveys published by the US Justice Department show that violent crimes, including rape, assault, robberies and murders, increased by 15 percent over the 1990s.
   d. Officially reported crime rates accurately reflect the actual incidence of crime.

9. Edwin Sutherland, who argued deviant behavior is behavior learned through interaction with others, developed which theory of deviance?
   a. social control theory
   b. social labeling theory
   c. alternative identity theory
   d. differential association theory

10. Ramiro Martinez's research on lethal violence in Latino communities indicates that _____.
    a. the likelihood of lethal violence in Latino communities is high, regardless of the socioeconomic conditions
    b. there is a clear link between the likelihood of lethal violence and the socioeconomic conditions for Latinos
    c. the likelihood of lethal violence in Latino communities is low, regardless of the socioeconomic conditions for Latinos
    d. none of these choices

11. According to the US Bureau of Justice Statistics, members of which group are most likely to be physically assaulted?
    a. White men
    b. White women
    c. Black men
    d. Black women

12. According to Robert Merton, deviance may occur when individuals _____.
    a. organize a social movement to raise awareness about a problematic behavior, redefine that behavior as deviant, and strengthen sanctions to punish the behavior
    b. are labeling by powerful social groups and sanctioned within impersonal bureaucracies
    c. accept the goals of the society but lack the means to legitimately achieve those goals
    d. break their allegiance to the common value system within society

13. Functionalist theory posits that _____.
    a. deviant behavior is related to people's social attachment to society
    b. social structure, not just individual motivation, produces deviance
    c. what appears to be dysfunctional behavior may be functional for society
    d. all of these choices

14. Durkheim suggested that which type of suicide results from the excessive regulation of individuals by social forces?
    a. anomic
    b. egoistic
    c. altruistic
    d. communal

15. Mark was born with cerebral palsy and is confined to a wheelchair. In other people's minds, Mark's condition overrides all other features of his identity, thereby creating what type of status?
    a. elite
    b. master
    c. primary
    d. differential

16. W. I. Thomas argued that juvenile delinquency was brought on by the social disorganization of slum life and urban industrialism, reflecting his assertion that _____.
    a. deviance is an emotionally immature response to the experience of poverty
    b. adolescents are not cognitively mature enough to manage stressful situations
    c. poor individuals are genetically predisposed to delinquent or criminal behavior
    d. deviance is a normal response to the social conditions in which people find themselves

17. Which of the following statements about race, class, and crime is (are) **true**?
    a. There is a relationship between unemployment, poverty, and crime.
    b. Law enforcement is concentrated in lower-income and minority areas, which leads to more frequent arrests of residents.
    c. African Americans and Hispanics are more than twice as likely to be arrested for a crime as are Whites.
    d. all of these choices

18. When a person undergoes a process of social transformation in which a new self-image and new public definition of the person as deviant emerges, the person has developed a(n) _____.
    a. primary identity
    b. deviant identity
    c. insider status
    d. victim role

19. Which of the following statements about rape is accurate?
    a. Rape is the most underreported crime.
    b. Criminologists explain the reported increase of rape in recent years as the result of a greater willingness to report it, and not an actual increase in the extent of rape.
    c. Victimization rates indicate that race is not a factor in who is raped.
    d. While young women are more likely to be raped than older women, middle and upper-class women are just as likely to be raped as poor women.

20. Which of the following factors accounts for the growth in the prison population in recent years?
    a. The rate of violent crime has been increasing.
    b. Mandatory sentencing has increased the number of inmates and the length of their sentences.
    c. Enforcement of drug offenses has increased.
    d. Both mandatory sentencing and increased enforcement of drug offenses explain the growth in the prison population.

21. Which of the following statements about the policing of minorities is accurate?
    a. Racial minorities are more likely than the rest of the population to be victims of excessive use of force by the police.
    b. Studies show that while most cases of police brutality involve minority victims, the involved officers receive heavy penalties for such behavior.
    c. Research on automotive searches carried out by state troopers on the New Jersey turnpike indicates that while 80 percent of the vehicles searched were driven by Blacks and Hispanics, the searches were justified by the high percentage of searches that turned up contraband and other crime evidence.
    d. all of these choices

22. Which of the following statements about women and crime is accurate?
    a. Women's participation in crime has increased so dramatically in the last decade that criminologists predict that women's crime rate will soon surpass men's.
    b. Women are more likely than men to be victimized by crime, especially violent crime.
    c. Women are more fearful of crime than are men.
    d. all of these choices

True-False Questions

1. Social control theory assumes there is a common value system within society and breaking one's allegiance to that value system is the main cause of deviant behavior.

   TRUE or FALSE

2. Research on juvenile offenders indicates that White and Black youths who are arrested are equally likely to be imprisoned.

   TRUE or FALSE

3. One of the primary reasons for the rapid increase in the prison population has been the introduction of minimum mandatory sentences for drug offenders.

   TRUE or FALSE

4. Once a deviant label has been assigned to a person, that person can recover their non-deviant identity fairly easily if they can prove that they did not engage in the behavior that led to being labeled.

   TRUE or FALSE

5. Functionalist theory views the dominant class as controlling societal resources and using its power to create values, belief systems, and laws that support that power.

   TRUE or FALSE

6. Labeling theorists argue that the official statistics concerning deviant behavior, such as suicide rates, accurately reflect the actual incidence of such behavior.

   TRUE or FALSE

7. Lower-income people and members of racial minority groups in the United States are more likely than higher-income and White people to be defined as mentally ill.

   TRUE or FALSE

8. Taking into consideration women of all races and social class backgrounds, it is White, middle class women who have the highest likelihood of being raped in the United States.

   TRUE or FALSE

9. Durkheim's analysis of suicide suggests that students on college campuses who experience intense feelings of loneliness and hopelessness are at risk for committing anomic suicide.

   TRUE or FALSE

10. The United States has the highest rate of imprisonment in the world.

    TRUE or FALSE

Fill-in-the-Blank Questions

1. Mark was born with cerebral palsy and is confined to a wheelchair. In other people's minds, Mark's condition overrides all other features of his identity, creating a _____ status.

2. Racial _____ refers to the common policing practice of using race as the sole criterion on which a person is stopped and detained for suspicion of criminal activity.

3. The FBI classifies larceny, auto theft, and arson as _____ crimes.

4. Deviant _____ maintain their own values, norms, and rewards for deviant behavior, thereby creating a worldview that solidifies the deviant identity of the members.

5. The process whereby deviant behavior is increasingly defined and treated as resulting from an illness is the _____ of deviance.

Essay Questions

1. Using the concepts of deviant identity, deviant career and deviant community, discuss how a person becomes a gang member.

2. Identify the three types of suicide discussed by Emile Durkheim, and use his analysis to explain why suicide rates are higher for men than for women.

3. Provide evidence to support or refute the following statement: "The American criminal justice system treats all people according to the neutral principles of law."

4. What social factors help explain why women's participation in crime has been increasing in the United States?

5. Do prisons deter crime or rehabilitate criminals? Support your response with evidence.

Solutions

PRACTICE TEST

Multiple Choice Questions

1. D, 153 Sociologists criticize the medicalization of deviance on the basis that it ignores the effects of social structure on the development of deviant behavior, focuses on the pathological state of mind of the deviant, and proposes to "cure" the "sick" mind through therapy or rehabilitation, rather than changing the social conditions that gave rise to the behavior.

2. D, 150 Body piercing, a behavior that violates customary norms, is a type of informal deviance. Behavior that violates laws is formal deviance.

3. A, 157-8 Conflict theory views crime as rooted in the economic organization of capitalist societies, which results in competition for social resources. Functionalist theory sees deviance as resulting from a lack of social integration and shared norms. Psychological theories focus on individual deficiencies or impairments as the primary cause of deviant behavior. Differential association theory, derived from symbolic interaction theory, views deviance as behavior one learns through interaction with others who support alternative values.

4. D, 167 Hate crimes are malicious acts against people or property that are motivated by bias based on race, ethnicity, religion, sexual orientation, or disability. Victimless crimes are violations of laws that have no complainant, such as prostitution. Organizational deviance is wrongdoing that occurs within the context of a formal organization and is sanctioned by the norms and operating principles of the organization. Elite deviance, also known as corporate crime, refers to the wrongdoing of wealthy and powerful individuals and organizations.

5. C, 166 Joel Best, as reported in the "Images of Violent Crime" box on p. 166, argues that violence is patterned according to the social characteristics of both perpetrators and victims. Violent crime is predictable, not random.

6. B, 154 Durkheim defined anomie as the condition that exists when social regulations break down and people exist in a state of relative normlessness. Deviance is behavior that violates norms, and crime is deviant behavior that violates criminal laws. Altruism refers to behavior that is devoted to the interests of others.

7. B, 155 According to Durkheim, egoistic suicide occurs when people become detached from society, as in the case of elderly men who have lost their work roles. Anomic suicide occurs when disintegrating forces in society make individuals feel intensely alone and hopeless. Altruistic suicide occurs due to the excessive regulation of individuals by social forces.

8. B, 165-7 The FBI's Uniform Crime Reports stress index crimes (violent personal crimes and property crimes), providing a relatively inflated picture for those crimes and underreporting elite crimes. Violent crimes peaked in 1990, but have decreased and leveled off since then. Besides the biases built into official crime statistics, they include only reported crimes, and thus do not accurately reflect the actual incidence of crime.

9. D, 159 Sutherland developed differential association theory, which emphasizes that deviance is learned through interaction with others. This theory developed out of symbolic interaction theory and the work of W.I. Thomas.

10. B, 169 Ramiro Martinez's research explored the connection between rates of violence in Latino communities and the degree of inequality in 111 U.S. cities. His research shows a clear link between the likelihood of lethal violence and the socioeconomics conditions for Latinos in these different cities.

11. D, 172 As indicated in Figure 6.5, Black women are most likely to be physically assaulted, followed by Black men, White men, and White women.

12. C, 155-6 According to Merton, deviance results from inconsistency between the goals and means of the society. For example, a prostitute accepts the goal of economic self-sufficiency supported by American values, but because she lacks the means to achieve this goal, she must develop alternative, albeit illegal, means.

13. D, 154 Functionalist theory posits that deviant behavior occurs when people's attachment to social bonds is weakened. Deviance, which results from structural strains in society, can be functional for society because it creates social cohesion and clarifies norms.

14. C, 154 According to Durkheim, altruistic suicide occurs due to the excessive regulation of individuals by social forces, as when an individual takes his or her life in the name of a greater cause.

15. B, 163 A physical disability may become a master status when that single characteristic overrides all other features of the person's identity. The disability may also be considered a stigma, or an attribute that is socially devalued and discredited.

16. D, 159 Using symbolic interaction theory as the basis for what he called situational analysis, Thomas examined people's actions and the subjective meanings they attributed to their behavior within their social context. Thomas argued deviance is a normal response to the social conditions in which people find themselves.

17. D, 169-72 There is evidence of a relationship between unemployment, poverty, and crime. Bias of official statistics, influence of powerful individuals, discrimination in patterns of arrest, and differential policing all affect the official rates of arrest and conviction. Bearing these factors in mind, however, there remains evidence that the actual commission of crime varies by race; racial minority groups are far more likely than Whites to be poor, unemployed, and living in single-parent families.

18. B, 160 Deviant identity is the definition a person has of herself or himself as deviant. The formation of a deviant identity involves self-acceptance of the label and role of "deviant." A deviant career is the sequence of movements a person makes through a deviant community, or a group organized around a form of deviant behavior, as they are socialized into the new role.

19. A, 172-3 Rape is the most underreported crime. Criminologists explain the reported increase of rape as the result of a greater willingness to report and an actual increase in the extent of rape. Poor and minority women are most likely to become a rape victim, as are young, single, divorced, or separated women.

20. D, 165-7 Despite the fact that the crime rate has been declining, prison populations have increased due to increased enforcement of drug offenses and mandatory sentencing.

21. A, 173 Most cases of police brutality involve minority citizens and there is usually no penalty for the officers involved. The New Jersey Turnpike research showed that the vast majority of the automobile searches carried out on vehicles driven by Blacks and Hispanics turned up no evidence of contraband or crimes of any sort.

22. C, 172-3 Although women's participation in crime has increased in recent years, women still commit fewer crimes than men. Women are less likely than men to be victimized by crime, although victimization varies significantly by race and age. Women are more fearful of crime than men.

True-False Questions

1. T, 156 Social control theory assumes there is a common value system in society and breaking one's allegiance to that system is the source of deviant behavior.

2. F, 171 Research on juvenile offenders found that 25% of arrested White youths are imprisoned, while nearly 60% of arrest Black youths are sent to prison.

3. T, 175 Increased enforcement of drug offenses and the introduction of minimum mandatory sentences for drug offenders have led to significant increases in the number of men and women imprisoned in the United States.

4. F, 159 According to labeling theory, it is very difficult for a person to recover their non-deviant identity once they have been labeled deviant, even if they did not engage in the behavior for which they were labeled.

5. F, 157-8 Conflict theorists view deviance as the result of competition for social resources within a capitalist system that is controlled by the dominant class. Functionalist theorists note that laws protect all people in the society, and economic interests alone cannot explain all the forms of deviance observed in society.

6. F, 161 Labeling theorists do not accept official statistics as accurate reflections of the actual incidence of deviant behavior, because these statistics are produced by institutions that define, classify, and record only certain behaviors as deviant. For example, one death may be classified as a suicide, while a similar death is classified as an accident.

7. T, 162 People with the fewest resources in society (including women, poor people, and members of racial-ethnic minority groups) suffer higher rates of reported mental illness and are most likely to be labeled mentally ill.

8. F, 173 Women who are most powerless in society (including poor women, young women, and racial-ethnic minority women) are more likely to be raped.

9. T, 154 College students who experience feelings of intense loneliness and hopelessness within the social context of relative normlessness may commit anomic suicide.

10. T, 174 As indicated in Figure 6.6, the United States has the highest rate of imprisonment in the world, followed closely by Russia.

Fill-in-the-Blank Questions

1. master, 163
2. profiling, 173
3. property, 167
4. communities, 161
5. medicalization, 153

Essay Questions

1. Deviant Career. Deviant careers, deviant identity and deviant communities are discussed on p. 160-161 of the text. Students applying these concepts to the process of a person becoming a gang member should explain how a deviant identity emerges, the stages of a deviant career, and the role of a deviant community.

2. Suicide. Emile Durkheim's three types of suicide are discussed on p. 154. Students using his analysis to explain why suicide rates are higher for men than for women should consider each of the three types of suicide and their relationship to social structure.

3. Equality Before the Law. The American criminal justice system, and its differential treatment of racial and ethnic minorities, is discussed on p. 169-174. Student responses should consider evidence for each aspect of the criminal justice system.

4. Women and Crime. The social factors explaining the increasing rate of women's participation in crime are discussed on p. 172. Student responses should demonstrate a knowledge of the social factors contributing to women's participation in crime, as well as ability to interpret crime statistics and trends.

5. Prisons: Deterrence or Rehabilitation. The question of whether prisons deter crime or rehabilitate criminals is discussed on p. 174-176. Student responses should demonstrate ability to work with statistics on crime and recidivism and to consider prisons in social context.

CHAPTER 7

SOCIAL CLASS AND SOCIAL STRATIFICATION

BRIEF CHAPTER OUTLINE

CHAPTER FOCUS

This chapter provides an overview of the social class system in the United States and explores the fundamental sociological questions of why inequality exists and what consequences inequality has for diverse groups and the society as a whole.

QUESTIONS TO GUIDE YOUR READING

1. What is social stratification and how do sociologists distinguish between the three stratification systems of estate, caste, and class?

2. How do the theories of Marx and Weber explain the existence of inequality?

3. How do the perspectives of functional and conflict theories on inequality differ, and what are the strengths and weaknesses of each perspective?

4. What are the layers of the social class structure in the United States, what is the distribution of wealth and income, and what are the sources of class conflict?

5. What issues are at stake in the race-class debate, and what are the strengths and weaknesses of each side of the argument?

6. What is the influence of gender and age on social stratification?

7. What is class-consciousness and how can false consciousness be explained?

8. What is social mobility and what is the extent of upward and downward mobility in the United States?

9. Who are the poor in the United States, and what are the causes of poverty?

10. How effective have welfare reform and the TANF program been in addressing poverty?

KEY TERMS

(defined at page number shown and in glossary)

caste system 183

class consciousness 199

educational attainment 189

false consciousness 199

ideology 185

life chances 184

occupational prestige 189

prestige 189

social differentiation 182

social stratification 182

status 182

urban underclass 191

class 182

culture of poverty 204

estate system 182

feminization of poverty 202

income 189

median income 189

poverty line 200

social class 184

KEY PEOPLE

(identified at page number shown)

CHAPTER OUTLINE

I. **SOCIAL DIFFERENTIATION AND SOCIAL STRATIFICATION**

   Status refers to a socially defined position in a group or a society. **Social differentiation** is the process by which different statuses in any group, organization, or society develop. **Social stratification** is a relatively fixed, hierarchical arrangement in society by which groups have different access to resources, power, and perceived social worth. All societies have a system of structured inequality; however, the basis on which groups are stratified varies cross-culturally.

   A. Estate, Caste, and Class
      1. In an **estate system** of stratification, which is most common in agricultural societies, the elite have total control over societal resources, as in the European feudal systems of the Middle Ages.
      2. In a **caste system** of stratification, one's status is ascribed, or assigned at birth. Examples include the former *apartheid* system in South Africa and the traditional caste system of India.
      3. In *class systems*, one's status is partially achieved, or determined by one's personal achievements. Although the class system of the United States is open, allowing for movement from one class to another, relatively few people experience much mobility over time. The social class a person is born into still has major consequences for that person's life.

   B. Defining Class
      1. **Social class** (or **class**) is the social structural position that groups hold relative to the economic, social, political, and cultural resources of society. *Class* is both an attribute of individuals and *a feature of society*.
      2. Max Weber identified **life chances** as the opportunities that people in a particular class have in common by virtue of membership in that class, including access to jobs, health care, housing, and education.
      3. Because social class is a structural phenomenon that cannot be directly observed, sociologists measure it by using indicators, or measurements that assess a concept, including income, education, occupation, and place of residence.

II. **WHY IS THERE INEQUALITY?**

   A. Karl Marx: Class and Capitalism

1. **Karl Marx** (1818-1883), who analyzed the class system under capitalism, defined classes in terms of their relationship to the *means of production*, or the system by which goods are produced and distributed.

2. Marx identified two primary social classes under capitalism -- the *capitalist class*, which owns the means of production, and the *working class*, or proletariat, which labors for wages.

3. Within the two major classes are two other classes -- the *petty bourgeoisie*, or small business owners and managers, who identify with the capitalists, and the *lumpenproletariat*, or those who become unnecessary as workers and are discarded. Today, this class includes the homeless and permanently poor.

4. Marx predicted that class struggles would occur because the two main classes would become increasingly polarized as wealth became more concentrated; thus, the exploitive character of capitalism would lead to its own destruction.

5. Capitalism provides the *infrastructure* of society and other social institutions reflect capitalist interests; for example, the family and education socialize people into appropriate work roles.

6. The dominant ideas of society are promoted by the ruling class, resulting in an **ideology** that supports the status quo.

7. Marx did not foresee the emergence of a large, highly differentiated, and culturally influential middle class like the one that exists in the U.S. today.

B. Max Weber: Class, Status, and Party

 1. **Max Weber** (1864-1920) agreed with Marx that social class has a powerful effect on people's lives, but argued that social stratification was *multidimensional*.

    a. *Class* (the economic dimension) refers to how much access an individual or group has to the material goods of society, and is measured by income, property, and other financial assets.

    b. *Status* (the cultural dimension) is the social judgment of, or recognition given to, a person or group.

    c. *Party*, or what we now call power (the political dimension), is the ability to influence people, even in the face of opposition. It is also reflected in one's ability to negotiate through social institutions, such as the criminal justice system.

 2. Weber noted that a person can rank high on one or two dimensions of stratification and low on another. For example, ministers are accorded high social status but do not typically earn high incomes.

C. Functionalism and Conflict Theory: The Continuing Debate

 1. *The Functionalist Perspective on Inequality.* Functionalist theory views society as an interdependent system of institutions organized to meet society's needs. It emphasizes cohesion and stability.

    a. Davis and Moore argue that inequality is a mechanism to ensure that the most talented people go into the most demanding positions.

    b. The higher rewards attached to these positions ensure that people will make the sacrifices needed to acquire the necessary training to succeed.

2. ***The Conflict Perspective on Inequality.*** Conflict theory emphasizes that society is a system held together through conflict and coercion, focusing on the friction in society rather than the coherence.

    a. Stratification is a system of domination and subordination whereby the unequal distribution of rewards reflect the class interests of the powerful, rather than the survival needs of the whole society.

    b. The more stratification in the society, the less likely the society will benefit from the talents of all its citizens because inequality limits the life chances of those at the bottom, who experience blocked opportunities.

3. Implicit in the argument presented by each perspective is criticism of the other. The fundamental, contradictory assumptions that drive each academic theory have practical implications for the development of social policy.

III. THE CLASS STRUCTURE OF THE UNITED STATES

The class structure of the United States is an elaborate, open system that can be conceptualized as a ladder. Social class is the common position that groups hold in a status hierarchy. **Social attainment** is the process whereby people end up in a particular position in the stratification system, influenced by such factors as class origins, educational level, and occupation. One's **socioeconomic status (SES)**, or an individual's position in the stratification system, is derived primarily from income, occupational prestige, and educational attainment. **Income** is the amount of money a person receives in a given period. The **median income** is the midpoint of all household incomes in a particular society [Figure 7.1]. **Prestige** is the value assigned to people and groups by others. **Occupational prestige** is the subjective evaluation people give to jobs as determined through nationwide surveys of the American public. Occupations such as judge, physician, professor, lawyer, and scientist are accorded the highest prestige, while electrician, newspaper columnist, insurance agent, and police officer are in the middle range of prestige ratings. Those occupations typically considered to have the lowest prestige are farm laborer, maid or servant, janitor, and garbage collector. These rankings reflect social judgements about the value of these jobs to society, rather than the worth of the people who hold these jobs. **Educational attainment** is typically measured as years of formal education. The amount of occupational prestige attributed to jobs is strongly related to the amount of education required by the job; thus, one's status generally increases as education level increases.

A. Layers of Social Class

1. The *upper class* in the United States constitute a very small proportion of people who control vast amounts of property and wealth, most of which is inherited. In Marx's terms, these elites own the means of production. The *nouveau riche*, such as owners of successful dot com companies, have moved into the upper class through the recent acquisition of wealth.

2. The *upper-middle class* includes those with high incomes and high social prestige, such as well-educated professionals and business executives.

3. The *middle class* is probably the largest group in the United States, or at least 45 percent of Americans identify themselves as middle class.

4. The *lower-middle class*, also known as the working class, includes lower-income bureaucratic workers, service workers, and workers in the skilled trades.

5. The *lower class* is comprised primarily of poor and displaced people who have little formal education and are often unemployed or working for minimum wage. People of color and women are over-represented in this class. Contrary to popular belief, forty percent of the poor work.

6. The **urban underclass** refers to people who are likely to be permanently unemployed and dependent on public assistance or crime for economic support. Sociologist William Julius Wilson argues that major structural changes in the economy have left racial minorities in especially vulnerable positions, thereby intensifying the problems of urban poverty.

B. Class Conflict
1. Rather than defining social class in terms of a hierarchy, or "ladder," conflict theorists define classes in terms of their structural relationship to each other and to the economic system.
2. This theory emphasizes power relations in society, noting that classes compete with each other for resources and the capitalist class exploits the working class.
3. The position of the middle class, or the *professional-managerial* class (including managers and professionals) is unique because members have substantial control over other people, especially in the workplace, but have minimal control over the economic system. This theory suggests that as capitalism progresses, members of the middle class will be pushed down into working class jobs.
4. Although the middle class has not been eliminated as Marx predicted, classes have become more polarized, and many members of the middle class have experienced downward social mobility due to economic restructuring.
5. Using Marx's definition, the working class would include many white-collar workers (such as secretaries, nurses, and salespeople) because they have little control over their work lives and little power to challenge the decisions of those who supervise them.
6. Because the middle and working classes shoulder much of the tax burden for social programs, they tend to develop resentment toward the poor, who are blamed for their own poverty. At the same time, privileges for the wealthy elite are perpetuated through declines in corporate taxes.

C. The Distribution of Wealth and Income
1. Class inequality in the United States is enormous and has intensified as elites have gained more power and greater control over wealth.
2. Sociologists distinguish between **wealth**, which is the monetary value of everything one owns (including stocks, bonds, property, insurance, and investments); and **income**, which is the amount of money brought into a household from various sources during a given period of time (including wages, investment income, and dividends).
3. The gap between the rich and the poor is increasing [Figure 7.2]. For example, median income (in constant dollars) has actually declined since 1970.
   a. The upper 20 percent and the upper 5 percent of the population have experienced the greatest group, while the income growth of everyone else has remained flat.
   b. The top 20 percent of the population has received a larger share of the total income (50% of all income), whereas the bottom 20 percent has received less (3.5%).
   c. Since 1980, the average pay of working people has increased by 74 percent, whereas the average pay of CEOs has increased 1884 percent.
   d. With the wealthiest 1 percent of the population owning 33 percent of all net worth, the concentration of wealth is higher in the United States than in any other industrialized nation.

e. With one-third of all Americans having a net worth of less than $10,000 once debt is subtracted, and debt continuing to increase in the typical American household, the American dream is increasingly unattainable.

4. At all income, occupational, and educational levels, Black families have lower levels of wealth than similarly situated White families. Because the advantages of wealth accumulate over time, providing equality of opportunity in the present has not mediated the consequences of years of discriminating against Black Americans in housing, lending, and other social policies.

5. Without significant wealth holdings, families of any race or ethnicity are less able to transmit assets from one generation to the next, which is one of the main supports for social mobility.

6. The new economy has mixed results: income levels for women have increased, but they have decreased for men, except for the top 30% of earners. Women in the bottom 10% of wage earners have also seen their wages decline. The tax structure has also distributed benefits unevenly, with corporate taxes having fallen, while American households are paying more in federal tax than ever before. Individuals at the upper ends of the class system can take advantage of numerous tax benefits and loopholes to reduce their tax burden.

7. Class divisions in the United States are real, apparently permanent, and becoming more marked as economic restructuring concentrates more wealth in fewer people's hands; downsizing redistributes formerly middle-class employees into jobs with lower pay, less prestige, and few (or no) benefits; and reductions in government programs eliminate much of the safety net for people in need.

D. Diverse Sources of Stratification

1. ***The Race-Class Debate.*** The class structure among African Americans has existed alongside the White class structure, separate and different.

   a. Both the African American and Latino middle class have expanded in recent years as a result of increased access to education and middle-class occupations for people of color. Their hold on middle-class status, however, is tenuous because of the persistence of racial discrimination and the vulnerability of their positions to periods of economic recession or political conservatism.

   b. The *myth of the model minority*, applied to Asian Americans, obscures the significant obstacles to success that Asian Americans encounter, ignores the hard work and educational achievements of other racial and ethnic groups, and obscures the high rates of poverty among many Asian American groups.

2. ***The Influence of Gender and Age***

   a. The measurement of class status for women is complicated by the question of how women's status is derived and the fact that circumstances can dramatically affect women's status. Class differences among women are also important. How each group of women experiences their gender depends on class, as well as race and position.

   b. Different ages groups experience quite different locations in the stratification system, and belonging to a particular generational cohort can have a significant influence on one's life chances. Today, the age group most likely to be poor in the U.S. is children, 17.6 percent of whom live in poverty; 10% of those 65 and older are poor.

IV. **SOCIAL MOBILITY: MYTHS AND REALITIES**

    A. Class Consciousness
        1. **Class consciousness** was higher in the U.S. during the labor movements of the 1920s and 1930s. Racial and ethnic differences in the working class, a relatively high standard of living, and a belief that anyone can succeed militate against class discontent.
        2. **False consciousness**, the extent to which subordinate classes have internalized the view of the dominant class, is also a factor.

    B. Defining Social Mobility
        1. **Social mobility** can be up or down, *intergenerational* or *intragenerational*, and *class systems* can be relatively *closed* or *open*.
        2. Despite the popular characterization of the U.S. class system as open, most people remain in the same class as their parents, and many actually drop. The cultural idols of social mobility are typically men.

    C. The Extent of Social Mobility.
        1. The limited mobility that exists in the U.S. is related to changes in the occupational system, economic cycles, and demographic factors.
        2. Upwardly mobile persons are often expected to distance themselves from their origins, which can result in many conflicts with family, with friends, and within themselves. The experience of upward mobility varies significantly by race and gender, and involves a collective effort of kin and sometimes the community. Even if one's class position changes, class origins continue to shape one's social experience.
        3. Doing better than one's parents has long been a hallmark of the American dream, but *downward mobility* is becoming more common and for the first time in U.S. history, the middle class is experiencing a decline.

V. **Poverty**

    A. Who Are the Poor?
        1. The vast majority of the 35.9 million people in the U.S. who live below the **poverty line** (12.5 percent of the population) are women and children.
        2. The **feminization of poverty**, which has been increasing in recent years, results from several factors including the growth of female-headed households (one-fourth of all families headed by women are poor), a decline in the proportion of the poor who are elderly, continuing wage inequality between women and men, and budget cutbacks in federal support programs and in federal employment.
        3. Inner cities became the home of most of the poor by the mid-1980s, and poverty rates are highest in the most racially segregated neighborhoods.
        4. Forty percent of the homeless population is families; 20 to 25 percent of the poor are mentally ill.

    B. Causes of Poverty.

        Two views about the causes of poverty prevail—that poverty is caused by the cultural habits of the poor or that poverty has structural roots.

1. ***Blaming the Victim: The Culture of Poverty.*** The **culture of poverty** argument views poverty as caused by welfare dependency, the absence of work values, and the irresponsibility of the poor. Research fails to support the contention that a culture of poverty is transmitted across generations. Children of poor parents have only a 16 to 28% probability of being poor adults, and many people classified as poor remain so for only one or two years. Most of the able-bodied poor work, with the working poor constituting 18 percent of the workforce. Of all poor persons, 25% work. The loss of benefits like healthcare coverage, wages too low to support a family, and workload costs like childcare make it difficult for the working poor to hold jobs and emerge from poverty.

2. ***Structural Causes of Poverty.*** The underlying causes of poverty lie in the economic and social transformations occurring in the U.S., the most important of which are the *restructuring of the economy* and *the status of women in the family and the labor market*.

C. Welfare and Social Policy.
1. The AFDC program created in 1935 as part of the Social Security Act was eliminated in 1996 by the Personal Responsibility and Work Reconciliation Act that administers welfare programs through TANF.
2. The law, implementing a policy known as *workfare*, stipulates a lifetime limit of five years for people to receive welfare, requires all welfare recipients to find work within two years, denies benefits to unmarried teen parents under 18 years of age unless they are in school and live with an adult, and requires unmarried mothers to identify the father.
3. Sociological research does not support the principles of the workfare policy. Higher level of welfare benefits can actually hasten exit from poverty, wages are often not sufficient to support a family, and daycare and other services are not adequate to support working parents.
4. Since welfare reform, there has been an increase in housing evictions, phone and utility cutoffs, and in some states an increase in both those neither working nor receiving welfare.
5. Sociologists conclude that the so-called welfare trap is not a matter of learned dependency, but a pattern forced on the poor by the requirements of sheer economic survival.

INTERNET EXERCISES

1. Browse through the U.S. Bureau of the Census website (www.census.gov/) and pick a topic related to diversity, income, housing, or poverty. Read and interpret the tables on your subject. Learn how to use the mapping function and credit a map for your state or the nation based on variables from your search. The American Community Surveys might also be useful for class projects.

2. The National Coalition for the Homeless website (nationalhomeless.org) has extensive resources in its On-line Library and Projects, which include a number of educational and campaign projects on issues like the educational rights of homeless children, voting, civil rights, healthcare and welfare monitoring. Select a topic and find related fact sheets, legislative information, and organizing tools to produce a report for sharing in class. Search for local housing organizations you can contact, for a research

paper or service-learning project. See also National Low Income housing Coalition (www.nlihc.org/) and the National Housing Trust Fund Campaign (www.nhtf.org/).

3. The Urban Institute (www.urban.org/) is a rich resource for research and policy analysis on a range of social and economic issues, including welfare and poverty policy, healthcare, prisoners and reentry, housing, immigration, at risk youth and more. Select a topic and search the site for a course project. You might want to start with the news bulletins, which feature developments you may not have noticed in the mainstream news.

INFOTRAC EXERCISES

Conduct a keyword search using InfoTrac College Edition to extend the discussion in DOING SOCIOLOGICAL RESEARCH: *The Fragile Middle Class* (p. 196-7).

1. **Keyword: Personal bankruptcy.** This search yields articles on the rates, types, reasons for, and distribution of personal bankruptcies.

2. **Keyword: Credit card debt.** This search leads to articles on college students, life style, debt education, and the economy.

3. **Keyword: Student debt.** There are several articles on soaring student debt, its effects on delaying family and home purchase, and its relation to choice of major and years in college.

4. **Keyword: Downward mobility.** Articles from this search link downward mobility to race and class, women, intergenerational mobility, refugees, the "sandwich" generation, and Latino workers.

5. **Keyword: Fragile middle class.** A search using this expression yields reviews of the book that is the subject of this feature.

PRACTICE TEST

Multiple Choice Questions

1. Which of the following statements about the estate system of stratification is true?
    a. The apartheid system of South Africa was a stark example of this type of system.
    b. Because there is little differentiation between classes in this type of system, considerably less social inequality exists than in other systems.
    c. One's status in this type of system is determined largely by one's personal achievements, allowing for social mobility within one's lifetime.
    d. Historically, this type of system was found in feudal societies where nobles controlled the land and peasants performed the labor.

2. Social differentiation refers to the _____.
   a. assignment of class position according to one's sex
   b. systematic inequalities between nations that result from differences in wealth, power, and prestige in the international economy
   c. process by which different statuses in any group or society develop based on which characteristics are deemed important in that society
   d. process whereby the most talented individuals, who perform the jobs that are most important in a society, receive the greatest rewards

3. According to Marx, which social class owns and controls the means of production?
   a. capitalists
   b. proleteriat
   c. petty bourgeoisie
   d. lumpenproletariat

4. According to Max Weber, the three most important dimensions of stratification in industrialized societies are _____.
   a. class, status, and party
   b. race, class, and gender
   c. power, party, and prestige
   d. economic, political, and religious

5. Which of the following factors have contributed to the feminization of poverty in recent years?
   a. an increase in the percentage of elderly people in poverty
   b. the persistence of wage discrimination for female workers
   c. an increase in the average number of children born to poor women
   d. all of these choices

6. In the United States, socioeconomic status (SES) is derived primarily from which three variables?
   a. income, occupation, and education
   b. occupation, wealth, and consciousness
   c. income, prestige, and political affiliation
   d. occupation, education, and religious affiliation

7. The degree of occupational prestige associated with a particular job is strongly correlated with which of the following variables?
   a. extent of opportunity for promotion in the job
   b. degree of danger involved in performing the job
   c. number of years of education required for the job
   d. degree to which the job makes a valuable contribution to the public good

8. Sociologists refer to those who become upper class through independent effort and have newly acquired wealth as the _____.
   a. capitalists
   b. Ivy League
   c. nouveau riche
   d. traditional elites

9. Sociologists refer to those people who are most likely to be permanently unemployed and dependent on public assistance or crime for economic support as the _____.
   a. lower-middle class
   b. working class
   c. surplus class
   d. urban underclass

10. According to Weber, life chances include the opportunity for which of the following things?
    a. marrying someone you love
    b. high self-esteem
    c. home ownership
    d. all of these choices

11. According to conflict theory, those who have substantial control over other people, especially in the workplace, but have minimal control over the economic system, are in which class?
    a. upper-elite
    b. working-laborer
    c. welfare-dependent
    d. professional-managerial

12. The amount of money brought into a household from various sources, such as wages, investments, and dividends, during a given period refers to a person's _____.
    a. debt
    b. wealth
    c. income
    d. surplus

13. The monetary value of everything a person owns, including property, stocks, bonds, and insurance, refers to a person's _____.
    a. debt
    b. wealth
    c. income
    d. surplus

14. Which of the following groups has recently experienced an expansion of the middle class due to increased opportunities for higher education and improved access to professional jobs?
    a. African Americans
    b. Native Americans
    c. Puerto Ricans
    d. none of these choices

15. In the United States, the age group most likely to be poor today is _____.
    a. elderly people over age 65
    b. children under age 18
    c. adults age 21 to 40
    d. adults age 41 to 60

16. The myth of the model minority, which is based on the assumption that a minority group member must adopt dominant group values to succeed, has been most frequently applied to which racial-ethnic minority group in the United States?
    a. Hispanic Americans
    b. African Americans
    c. Native Americans
    d. Asian Americans

17. When an individual moves from one class to another in his or her own lifetime, this is referred to as _____.
    a. generational fortune
    b. professional success
    c. intragenerational mobility
    d. intergenerational mobility

18. According to which theory does inequality serve an important purpose in society by motivating people to acquire the education and job training to fill positions necessary for society's survival?
    a. conflict
    b. feminist
    c. functionalist
    d. symbolic interaction

19. Which of the following factor(s) tend to increase class consciousness?
    a. strong cultural belief that upward mobility is possible
    b. the formation of a highly educated and relatively large middle class
    c. a relatively high standard of living
    d. none of these choices

20. Which of the following statements about the distribution of wealth and income in the U.S. is (are) true?
    a. Since the early 1970s, the top 20% of the population has received a larger share of the total income, whereas the bottom 20% has received less.
    b. The average CEO in the U.S. now makes 475 times the earnings of an average worker.
    c. The wealthiest 1% own 33% of all net worth; the bottom 80% control only 16%.
    d. all of these choices

21. According to the culture of poverty argument, the major causes of poverty are _____.
    a. unemployment, lack of access to higher education, and high housing costs
    b. welfare dependency, absence of work values, and the irresponsibility of the poor
    c. economic restructuring, the low status of women in the family and the labor market, and racism
    d. disappearance of manufacturing jobs, declining wage rates, and the high costs of healthcare and childcare

22. Measured by their own incomes and occupations, rather than by their husbands', a majority of women in the United States would be members of which social class?
    a. working
    b. middle
    c. upper
    d. under

23. The fastest growing segment of the homeless population in the United States is _____.
    a. single adult women
    b. runaway adolescents
    c. families with children
    d. men who are mentally ill

24. Which of the following changes to the American welfare system were instituted under the Personal Responsibility and Work Reconciliation Act adopted in 1996?
    a. All recipients were limited to a lifetime benefit limit of 2 years.
    b. Teenage recipients were required to stay in school and live with an adult while receiving welfare benefits.
    c. Recipients were required to place their children in foster care if they could not financially support them without government assistance.
    d. all of these choices

25. Which of the following statements about the poor in the United States is true?
    a. Over half of all families headed by women are poor.
    b. About half of the poor live inside central cities.
    c. The majority of the poor live in isolated rural areas.
    d. The majority of the poor receive food stamps, housing assistance, and Medicaid.

True-False Questions

1. In the United States, women have significantly higher rates of poverty than do men.

   TRUE or FALSE

2. Despite beliefs to the contrary, class divisions in the United States are becoming less marked, with the positions of the middle class, working class, and poor improving in the last two decades.

   TRUE or FALSE

3. The estate system of stratification is found primarily in agricultural societies.

   TRUE or FALSE

4. According to Max Weber, a person's position in the stratification system is almost solely determined by their income.

   TRUE or FALSE

5. The majority of upper class people in the United States acquire their wealth through independent efforts, such as establishing a successful business.

   TRUE or FALSE

6. As adults, most individuals in the United States remain in the same social class as their parents.

   TRUE or FALSE

7. Despite the development of a strong Black middle class, at all levels of income, occupation, and education, Black families have lower levels of wealth than similarly situated White families.

   TRUE or FALSE

8. Most Americans are paying less federal tax today than ever before because of changes in the tax system that have placed a greater tax burden on corporations.

   TRUE or FALSE

9. "Median income" refers to the minimum amount of income that a family needs to earn to support their basic needs, such as housing, food, and clothing.

   TRUE or FALSE

10. Since the implementation of TANF, welfare rolls have shrunk, and poverty rates have declined significantly as former recipients find jobs.

    TRUE or FALSE

Fill-in-the-Blank Questions

1. According to Marx, _____ consciousness develops when subordinate classes internalize the views and belief systems of the dominant class, which justify inequality.

2. According to _____ theory, social inequality ensures that the most talented people go into the most demanding positions in society by rewarding them for investing in education and training.

3. The United States is a(n) _____ class system because social mobility is permitted and an individual's placement in the system may change over time.

4. In 2003, the official _____ in the United States was $18,725 for a family of four people.

5. In a _____ system of stratification, one's position is determined through ascribed status, as in the former apartheid system in South Africa.

Essay Questions

1. Define life chances and explain how the social class position of one's family of origin contributes to one's life chances in adulthood.

2. Explain why there are differences in wealth between Black and White Americans at all educational, income, and occupational levels.

3. Identify three factors that have contributed to the trend known as the feminization of poverty, and suggest one possible solution for addressing this problem.

4. Discuss why the myth of social mobility persists in the United States despite considerable evidence to the contrary.

5. Explain why the new welfare policy (TANF) has not reduced poverty.

Solutions

PRACTICE TEST

Multiple Choice Questions

1. D, 182 The estate system of stratification is usually found in agricultural societies. It is characterized by stark inequality between those who own the land (elites) and those who work on it (commoners or peasants). The estate system characterized the feudal societies of Europe during the Middle Ages.

2. C, 182-3 Social differentiation is the process by which different statuses in a group or society develop. Social stratification refers to the relatively fixed, hierarchical arrangement in society that distributes social resources unequally across groups.

3. A, 185 According to Marx, the two primary classes under capitalism are the capitalists, who own and control the means of production, and the working class, or proletariat, who sell their labor. The petty bourgeoisie are the skilled craftspeople who identify with the capitalist class and the lumpenproletariat are the group of dispossessed workers.

4. A, 186 Weber identified three important dimensions of stratification: economic (class or income), cultural (status or prestige), and political (party or power).

5. B, 202 The persistence of wage discrimination, high rates of divorce and the associated lack of men's financial contributions to the household, and reductions in government assistance programs have intensified the feminization of poverty.

6. A, 189 In the United States, socioeconomic status is determined by three indicators – income, education, and occupation.

7. C, 189 Occupations that require high levels of education, such as judge, physician, and scientist, are accorded high occupational prestige. Jobs such as maid and laborer are typically associated with the least prestige. Nurses and police officers are in the middle range.

8. C, 190-1 The nouveau riche are people who have recently acquired substantial wealth through independent effort, rather than inheritance, including the owners of successful dot com companies.

9. D, 191 The urban underclass includes those people who experience chronic unemployment and are likely to be dependent on government assistance or crime for economic support. Marx identified disposed workers like these as the lumpenproletariat.

10. C, 184 Weber identified life chances as the opportunities that people have in common by virtue of belonging to a particular class. Life chances include the opportunity for having a certain income, possessing goods, and having access to particular jobs.

11. D, 192 According to conflict theory, the professional-managerial class has substantial control over other people but minimal control over the economic system, placing them in a position of identifying with the capitalist class yet laboring for wages.

12. C, 189 Income is the amount of money brought into a household in a given period.

13. B, 189 Wealth refers to the total monetary value of everything one owns. Net worth refers to a person's or a family's total assets minus their debts.

14. A, 197 Both Latinos and African Americans have experienced an expansion of the middle class through higher education and increased access to middle-class jobs; however, many middle-class African Americans have a tenuous hold on their class position. Puerto Ricans have the highest rate of poverty among Latinos, and Native Americans have the highest poverty rate of all groups in the United States.

15. B, 199 Older people were most likely to be poor in the past, but today, children are the age group most likely to be poor. Some 17.6 percent of American children live in poverty.

16. D, 197 Asian Americans are often referred to as the "model minority;" however, many Asian American subgroups experience high rates of poverty, and even those groups that have assimilated into the dominant culture continue to experience prejudice and discrimination.

17. C, 200 Intragenerational mobility occurs when an individual's class status changes over his or her lifetime. Intergenerational mobility occurs when an individual's class changes from his or her parents' class. Social mobility, which may be upward or downward, occurs less frequently than is commonly believed.

18. C, 187 Functionalist theorists view upward mobility as available to those who acquire education and job skills, whereas conflict theorists argue mobility is blocked for the lower and working classes, who do not have the same opportunities and resources.

19. D, 199 A strong cultural belief in the possibility of upward mobility, a highly educated and relatively large middle class,

and a relatively high standard of living are all factors that tend to decrease class consciousness.

20. D, 193 Since the early 1970s, the income of the top 20% of the population has increased to 50% of all income, whereas the income of the bottom 20% has decreased to only 3.5%. The average CEO in the U.S. now makes 475 times the earnings of the average worker. The wealthiest 1 percent own 33% of all net worth, whereas the bottom 80% control only 16%

21. D, 204 The culture of poverty thesis ignores structural causes of poverty and essentially blames the poor for their plight.

22. A, 198 Most women in the United States would be considered working class by their own incomes and occupations, because women are disproportionately located in low-status, low-wage jobs, despite having educational levels comparable to men.

23. C, 203 Families with children are the fastest growing segment of the homeless population in the United States; families are now 40% of the homeless.

24. B, 205-6 The Personal Responsibility and Work Reconciliation Act placed a 5 year limit on receiving benefits and required that recipients secure a job within 2 years or agree to do community service work without being paid. The Act also requires teen mothers who receive benefits to stay in school and live in another adult's household. Additionally, women who refuse to identify the fathers of their children risk losing their benefits regardless of their eligibility for assistance.

25. B, 202 About one quarter of all families headed by women are poor, and about half of the poor live in central cities, and 9% in suburbs, with the remainder in rural areas. About 12.5 percent of the total population of the United States is poor. Almost two-thirds of the poor receive no food stamps; 80 percent get no housing assistance; and less than half receive Medicaid.

True-False Questions

1. T, 202 Women and children have always had higher rates of poverty than men, but this disparity has been intensified in the United States through the process known as the feminization of poverty.

2. F, 195 Class divisions in the United States are becoming more marked, with the positions of the middle class, working class, and

poor deteriorating in the last two decades.

3. T, 182 Estate systems of stratification are most common in agricultural societies.

4. F, 185-6 Marx emphasized the importance of class for one's placement in the class system, arguing that it was determined solely by one's relationship to the means of production (as owner or laborer). Weber identified three

dimensions as important—class, status, and party.

5. F, 190 Most of the wealth owned by the majority of the upper class is inherited. The nouveau riche have earned their wealth through personal achievement.

6. T, 200 The majority of people in the United States occupy the same class position as their parents.

7. T, 195 At all education, income, and occupation levels, Black families have lower levels of wealth than similarly situated White families.

8. F, 195 The average American pays more federal tax today than ever before, while the tax burden on corporations has been greatly reduced.

9. F, 200-1 The median income is the midpoint of all household incomes in a society. The poverty line, which was $18,725 for a family of four in 2003, is the minimal amount of income needed to support the household members' basic needs.

10. F, 207 Those who have entered the labor force most often earn wages that keep them below the poverty line. Other indicators like utility and phone cutoffs, and an increase in housing evictions also indicate an increase in poverty.

Fill-in-the-Blank Questions

1. false, 199
2. functionalist, 187
3. open, 200
4. poverty line, 200-1
5. caste, 183

Essay Questions

1. Life Chances. Life chances are defined on p. 184, and social mobility and wealth are also discussed on the same page of the text. Student responses should demonstrate an understanding of how life chances are related to social mobility and wealth.

2. Race and Wealth. The relationship between race and wealth is discussed on p. 195 of the text. Student responses should demonstrate an understanding of the impact of race on the acquisition of wealth, as well as on educational, income and occupational attainment.

3. Feminization of Poverty. The feminization of poverty is discussed on p. 202 of the text. Student responses should demonstrate understanding of the three factors that have contributed to the trend.

4. Myth of Social Mobility. The myth of social mobility is discussed on p. 199 of the text. Student responses should demonstrate an understanding of the myth and why it persists.

5. TANF and Poverty. The impact of the workfare policy of TANF is discussed on p. 205-207. Student responses should demonstrate a grasp of the main provisions of TANF and why these provisions have not successfully addressed the structural causes of poverty.

CHAPTER 8

GLOBAL STRATIFICATION

BRIEF CHAPTER OUTLINE

Global Stratification

Rich and Poor

The Core and Periphery

Race and Global Inequality

Theories of Global Stratification

Modernization Theory

Dependency Theory

World Systems Theory

Consequences of Global Stratification

Population

Health and Environment

Education and Illiteracy

Gender Inequality

War and Terrorism

World Poverty

Who Are the World's Poor?

Women and Children in Poverty

Poverty and Hunger

Causes of World Poverty

Globalization and Social Change

Chapter Summary

CHAPTER FOCUS

This chapter examines the structure of global stratification, the theories explaining it, its consequences, and the interconnectedness it brings through the international division of labor. It also explores the patterns of population, health and environment, education and illiteracy, gender inequality, war and terrorism, and world poverty.

QUESTIONS TO GUIDE YOUR READING

1. What is the structure of global stratification and what are the various systems for defining global stratification?

2. What are the differences between the first, second, and third world country system for defining global stratification and the core, semi-peripheral, and peripheral country system?

3. How do modernization, dependency, and world systems theories explain global stratification, and what are the strengths and weaknesses of each theory?

4. What are the consequences of global stratification in terms of population, health, education, and gender?

5. What is a commodity chain, and what is its relation to the international division of labor?

6. What is world poverty, who are the world's poor, and what are the causes of world poverty?

7. What is the gender development index, and what does it indicate about the global status of women?

8. What conditions distinguish the newly industrializing countries from those that are not making it?

KEY TERMS

(defined at page number shown and in glossary)

absolute poverty 228

commodity chain 223

dependency theory 221

first-world countries 218

global stratification 214

human poverty index 228

modernization theory 220

neocolonialism 222

peripheral countries 218

relative poverty 228

semiperipheral countries 218

third-world countries 218

world systems theory 222

colonialism 221

core countries 218

extreme poverty 228

gender development index 226

gross national income (GNI) 215

international division of labor 219

multinational corporations 222

newly industrializing countries (NICs) 232

power 218

second-world countries 218

terrorism 228

CHAPTER OUTLINE

I. GLOBAL STRATIFICATION

Worldwide, there are not only rich and poor individuals, but there are also rich and poor countries. There is a system of **global stratification** in which units are countries; thus, nations cannot be seen independently of economic and social processes that link them together in a world-based economy. Several measures of well being, such as life expectancy and infant mortality, illustrate great inequities resulting from stratification.

A. Rich and Poor

Using annual **per capita Gross National Income (GNI)** as a measure of global stratification, we can compare the 10 richest and 10 poorest nations. The richest are the industrialized, mostly urban countries, primarily in Western Europe and including the United States and Japan, and the poorest are becoming industrialized, are largely rural and heavily dependent on subsistence agriculture, and are mostly in eastern or central Africa. Inequality between the rich and poor nations is marked and increasing.

B. The Core and Periphery
1. The countries of the world can be divided into three levels based on their **power** in the world economic system—the **core countries** (Europe, the United States, Australia, and Japan); the **semi-peripheral countries** (like Spain, Turkey, and Mexico); and the **peripheral countries** (the poor, largely agricultural countries).
2. During the Cold War, the categorization of **first-world**, **second-world**, and **third-world countries** reflected the political and economic dimensions of global stratification.

C. Race and Global Inequality
1. Countries with high standards of living, high levels of education, and other positive characteristics have populations that are predominantly White, while those countries with lower standards of living are predominantly people of color.
2. This new racial divide is based on the new **international division of labor** that relies on cheap labor that can be found mostly in non-Western countries.
3. Looking at racial categorization globally, Anthony Marx ironically suggests that the creation of arbitrary racial labels (as in the U.S. and South Africa) may in the long run lead to more racial equality than in countries lacking such racial categories (as in Brazil) by providing an identity around which political mobilization of people of color can take place.

II. THEORIES OF GLOBAL STRATIFICATION

A. Modernization Theory
1. **Modernization theory** views the economic development of a country as a process whereby traditional societies become more complex and differentiated by changing their attitudes, values, and institutions.

2. Based largely on functionalist theory and the work of Max Weber, this perspective argues that the Industrial Revolution occurred in Northern Europe because the Protestant residents there were hardworking people who valued thrift and individual achievement.

3. Similar to the "culture of poverty" theory, modernization theory views countries as poor because they have poor attitudes and poor institutions; thus, proponents of this theory recommend development as the solution to poverty.

4. Modernization theory has been criticized for not explaining the development, or lack thereof, in all nations; blaming countries for being poor; and arguing that government should not make economic decisions or develop policies that restrict business or free trade.

B. Dependency Theory

1. Derived from the work of Karl Marx, **dependency theory** argues that the poverty of low-income countries is a result of **colonialism** and economic exploitation by powerful countries.

2. **Neocolonialism** is a form of international control of poor countries by rich countries, without direct political or military involvement. In this process, rich industrialized nations set prices for raw materials produced by poor countries at very low levels so that the poor countries are unable to accumulate enough profit to industrialize.

3. **Multinational corporations**, who buy resources and labor in the countries with the lowest prices, play an important role in keeping dependent nations poor.

4. Critics of dependency theory argue that some colonies, such as Hong Kong, have done well economically. Critics also emphasize that it is not clear that the involvement of multinational corporations always impoverishes nations.

C. World Systems Theory

1. **World systems theory** argues that there is a world economic system that must be understood as a single unit, not in terms of individual countries.

2. This theory, derived from conflict theory and most closely associated with the work of Immanuel Wallerstein, argues that a global system of stratification has developed based on historical and strategic imbalances in the economic system.

3. This theory divides the world into three groups of interrelated nations.
   a. *Core countries* are the rich, powerful, capitalistic countries that control the world system.
   b. *Semiperipheral countries* occupy an intermediate position in the world.
   c. *Peripheral countries* are poor, largely agricultural, and manipulated by core countries that extract resources and profits from them.

4. This focuses on the economic interconnectedness of countries worldwide and an international division of labor, which contributes to the global production of goods through a **commodity chain**, or the network of production and labor processes by which a product becomes a finished commodity.

5. The growing phenomenon of international migration is the result of refugees seeking asylum as well as the demand for cheap labor, as *transnational communities* within **world cities** are linked to international commerce.

6. Critics of world systems theory note that it is not clear that the world system always works to the advantage of core countries and to the detriment of peripheral countries, and furthermore, low-wage sweatshops are found in *all* nations, including core countries such as the United States.

III. CONSEQUENCES OF GLOBAL STRATIFICATION

A. Population
1. The poorest countries, comprising over 3 billion people (over half the world's population), have the highest birth rates and the highest death rates.
2. The richest countries have only 15 percent of the world's population, and many of these countries are experiencing population declines.
3. Scholars believe that while in some situations large population and high birth rates can impede economic development, in general *fertility rates* decrease and population growth levels off as countries develop.

B. Health and Environment
1. High-income countries (15% of the world's population) have lower childhood death rates, higher life expectancies, and fewer children born underweight; people also have access to clean water and adequate sanitation. Together they use more than half of the world's energy.
2. In poor countries, childhood death rates are high, life expectancy is considerably shorter, and fewer people have access to clean water and adequate sanitation.
3. Degradation of the environment affects all nations; deforestation is most severe in Latin America, Africa, Mexico and Southeast Asia, whereas overproduction of "greenhouse gas" is most severe in the U.S., Canada, Australia, parts of Western Europe and Russia where energy use per capita is highest.

C. Education and Illiteracy
1. In high-income countries, literacy and school attendance are taken for granted, and education is nearly universal.
2. The percent of elementary-age children in school is significantly lower in middle and lower income nations; however, even in poor areas, education is improving.
3. Most education around the world, including basic literacy and math skills, occurs in family settings or religious congregations. This informal education does not give most people the skills and knowledge needed to operate successfully in an increasingly technological world.

D. Gender Inequality

The **gender development index (GDI)** is calculated based on gender inequalities in life expectancy, educational attainment, and income in each country. In every nation, the GDI is less than the general HDI. Many countries have shown improvement in the GDI recently, including both industrialized and developing countries, indicating that gender equality can be achieved at different income levels and in different stages of national development.

E. War and Terrorism

By generating inequities in the distribution of power, global stratification results in international conflicts bring war and increased risk of **terrorism**. Global inequities and the dominance of Western culture and Western nations over others contribute to the roots of terrorism.

IV. WORLD POVERTY

Some measures of poverty focus on the amount of money available for basic survival. The United Nations has established the *international poverty line* as the situation in which individuals live on less than $365 a year, or in **absolute poverty**. **Extreme poverty** is defined as the situation in which people live on less than $275 a year. **Relative poverty** compares households in poverty with other households (as in the U.S.) The UNDP **human poverty index** to measure the degree of deprivation in four dimensions: long and healthy life, knowledge, economic well-being, and social inclusion, with different indicators used for developing and industrialized countries.

A. Who are the World's Poor?

Using the World Bank's definition of international poverty as consumption level below $1 per day, 28% of the world's population lives below the *international poverty line*. Poverty has increased in eastern Europe and central Asia in the last decade, but has decreased in east Asia, particularly in the People's Republic of China. About 42% of the population of sub-Saharan Africa lives in poverty with high infant mortality and low life expectancy, aggravated by AIDS.

B. Women and Children in Poverty
1. Women bear a larger share of the burden of world poverty, a situation called *double deprivation*. Women, who make up 60% of the world's population, perform two thirds of all working hours, receive only one-tenth of the world's income, and own less than one percent of the world's wealth. In poor countries, women also suffer greater health risks because of high fertility rates, risky childbirth, and traditions and cultural norms depriving them of adequate nutrition.
2. Children in poverty do not have the luxury of a childhood or education, and often work in sweatshops and even prostitution. The UN estimates that 211 million children (5-14 years old) are in the paid labor force, most in Asia and sub-Saharan Africa. An estimated 13 million homeless street children in Latin America survive through a combination of begging, selling, prostitution, drugs, and stealing.

C. Poverty and Hunger

An estimated 1.2 billion people in the world are so poor that they are unable to obtain enough food to meet their nutritional needs, and an estimated 5 million children under age 5 die each year in the developing world from malnutrition. The crucial issue is the proper distribution of food. Most areas of the world experienced a marked decrease in hunger since 1970, except for Sub-Saharan Africa where hunger has increased.

D. Causes of World Poverty

Poverty is increasing in areas with a history of unstable governments and where economies have collapsed, forcing governments to borrow under harsh economic restructuring plans set by the World Bank and the IMF. Poverty is also caused by changes in the world economic system, which have resulted in a sharp drop in commodity prices. Although the drop in prices benefited some newly industrializing Asian countries by reducing the cost of raw materials, commodity-producing nations in Africa and Latin America suffered economic collapse, leading to massive poverty and starvation.

V. GLOBALIZATION AND SOCIAL CHANGE

In the **newly industrializing countries (NICs)** in East Asia and parts of Latin America economic development is occurring, but some nations are not making it. Mass genocide and refugee migrations have increased poverty and hunger. Some of the newly created states of the former Soviet Union face enormous economic and social problems. Globalization has created great progress in the world, but it is also contributing to the inequality between nations and to the exploitation of some nations and groups by other, fomenting world conflict.

INTERNET EXERCISES

1. The United Nations website (www.un.org/) is an incredible network of documents and sites related to UN agencies. To research issues related to global stratification, the best route is through the section on Economic and Social Development. Under that site are several subject areas—like environment, human settlements, population, human rights, social development, sustainable development, and women—which are all UN agencies under the Economic and Social Council. Explore the site that most interests you, choose a project you consider innovative and interesting, and write a brief report on it. The UNDP issues an annual report featuring one issue related to development and proposed solutions, which might be relevant to your interests. See Women Watch, (www.un.org/womenwatch/) a UN website that brings together all UN agencies and information related to women.

2. The United Nations Children's Fund (www.unicef.org/) issues an annual State of the World's Children Report, which focuses on poverty and quality of life—a lens through which to consider global stratification.

3. The World Bank Group website (www.worldbank.org/) can be used for regional, country-by-country, or world development data for a research project focused on global stratification. You could also explore particular topics, like gender, and write a report on a model project or issue addressed by the World Bank.

INFOTRAC EXERCISES

Conduct a keyword search using InfoTrac College Edition to extend the discussion in DOING SOCIOLOGICAL RESEARCH: *Servants of Globalization: Who Does the Domestic Work?* (p. 219).

1. **Keyword: Migrant domestic workers.** This search yields articles on transnationalism and domestic workers in different global regions.

2. **Keyword: Global domestic labor.** This search leads to articles linking global domestic labor to gender, the urban-rural income gap, trafficking in women, the environmental and ethical aspects of international migration, and more.

3. **Keyword: Social policy and care work.** This search leads to articles on elder and child care, work and family, health, and women.

4. **Keyword: Housework.** This search yields articles on the gender division of household labor, changes in distribution of housework, the effects of housework on health, and the question of wages for housework.

PRACTICE TEST

Multiple Choice Questions

1. Which of the following is a measure of the total volume of goods and services produced by a country each year?
   a. annual commodity product
   b. annual human poverty index
   c. annual global economic index
   d. annual per capita gross national income

2. The majority of the wealthy countries are located in which part of the world?
   a. Western Europe
   b. Eastern Europe
   c. South America
   d. South Asia

3. The classification scheme used during the Cold War categorized the United States as a _____.
   a. peripheral country
   b. first-world country
   c. second-world country
   d. semiperipheral country

4. Which of the following features are characteristic of peripheral countries?
   a. They are largely urbanized.
   b. They are semi-industrialized.
   c. They have high poverty rates, despite often having valuable natural resources.
   d. all of these choices

5. Using power as the main dimension of stratification in the world system, the semi-industrialized nations of Spain and Turkey would be classified as _____ countries.
   a. core
   b. modern
   c. peripheral
   d. semiperipheral

6. Those countries that control and profit most from the world system, including the United States, Australia, and Japan, are classified as _____ countries.
   a. semiperipheral
   b. traditional
   c. peripheral
   d. core

7. The rapid expansion of the global capital system has resulted in a new international division of labor that is characterized by _____.
   a. an increase in racial inequality between nations
   b. the exploitation of cheap labor in non-Western countries
   c. the exodus of unskilled workers from poor countries into industrialized countries
   d. all of these choices

8. Which of the following nations uses the highest percentage of global energy?
   a. Japan
   b. Russian Federation
   c. China
   d. United States

9. In which of the following poor regions of the world has poverty been reduced since 1990?
   a. East Asia
   b. South Asia
   c. Eastern Europe
   d. Sub-Saharan Africa

10. Which of the following is a dimension of the gender development index?
    a. level of educational attainment
    b. rate of home ownership
    c. family size
    d. all of these choices

11. Modernization theory sees economic development as arising from the _____.
    a. economic interconnection of countries worldwide and the emergence of an international division of labor whereby products are produced globally
    b. role of neocolonialism in controlling and exploiting poor countries to prevent them from accumulating enough wealth to become industrialized
    c. adoption of new technologies and market-driven attitudes that promote saving and investing in traditional societies
    d. contribution of multinational corporations to maintaining poverty in dependent nations by keeping wages low

12. Based on the work of Karl Marx, _____ theory asserts that the poverty of low-income countries is a result of colonization and economic exploitation by rich, powerful countries.
    a. dependency
    b. modernization
    c. world systems
    d. global production

13. Which of the following is a dimension of deprivation used to calculate the human poverty index?
    a. economic well-being
    b. life expectancy
    c. social inclusion
    d. all of these choices

14. The United Nations identifies the situation in which individuals live on less than $275 per year as _____.
    a. global stratification
    b. double deprivation
    c. extreme poverty
    d. chronic poverty

15. Which of the following is a major cause of world poverty?
    a. The high cost of international loans and the harsh economic restructuring plans imposed by the World Bank and the IMF
    b. Rapid population growth
    c. Lazy people who are uninterested in working
    d. none of these choices

16. In the United States, the poverty level is determined by the annual income for a family of four that is considered necessary to maintain a suitable standard of living, which reflects ____ poverty.
    a. chronic
    b. relative
    c. absolute
    d. extreme

17. Which of the following nations would be considered a *newly industrializing country* based on its recent, rapid economic growth?
    a. United States
    b. Australia
    c. Africa
    d. South Korea

18. Women in poor countries are said to experience *double deprivation* because they _____.
    a. perform only one-third of all working hours and suffer from malnutrition
    b. constitute half of the world's population, but receive only one-quarter of the world's income
    c. suffer because of their gender and disproportionately carry the burden of poverty
    d. suffer from malnutrition and low fertility rates

19. The richest countries in the world have approximately what percent of the world's population?
    a. 15
    b. 25
    c. 50
    d. 75

20. Anthony Marx argues that the lack of clearly defined racial categories in Brazil has _____.
    a. resulted in a society where racial differences have no influence on individual opportunities or social status
    b. upset the elite class, who has repeatedly demanded that the government establish an official racial classification system
    c. denied dark-skinned people the opportunity to develop a strong group identity that would serve as the basis for group solidarity
    d. encouraged dark-skinned people to collectively pressure the government to recognize them as a distinct social group with a unique cultural history

True-False Questions

1. With the exception of Russia, the United States has a higher poverty rate than all other industrialized nations in the world.

   TRUE or FALSE

2. Most of the basic education provided to people in poor countries occurs in informal settings such as the family and religious congregations.

   TRUE or FALSE

3. According to the United Nation's definition of absolute poverty, over one-fourth of the world's population lives in poverty.

   TRUE or FALSE

4. Modernization theory, which is derived from conflict theory, views traditional societies as poor because they have been colonized and otherwise exploited by rich countries.

   TRUE or FALSE

5. World systems theory argues that the poverty of low-income countries is a result of having poor institutions and poorly motivated individuals who do not save or invest their money.

   TRUE or FALSE

6. Dependency theory classifies nations based on their degree of control over the global economic system.

   TRUE or FALSE

7. The populations of the poorest countries in the world are largely rural and comprised mainly of people of color.

   TRUE or FALSE

8. Both deforestation and "greenhouse gas" emissions are most severe in South America, Africa, Mexico, and Southeast Asia.

   TRUE or FALSE

9. In low-income countries, high childhood death rates are attributable to poor sanitation and contaminated water supplies as well as food shortages.

   TRUE or FALSE

10. One of the main problems facing rich countries today is that declining birth rates have led to a shortage of workers.

   TRUE or FALSE

Fill-in-the-Blank Questions

1. The form of international control whereby rich countries lend money to poor countries, resulting in the greater dependence of poor countries on rich countries, is called _____ .

2. Companies such as Nike are _____ , or large companies whose stockholders are from the industrialized countries, but whose materials are purchased in countries where resources are cheapest and whose goods are manufactured in countries where labor costs are lowest.

3. World systems theorists call the global network of production and labor processes by which a product becomes a saleable item the _____ .

4. Cities that are closely linked through the system of international commerce are called _____ .

5. According to the United Nations' definition, _____ poverty is the situation in which individuals live on less than $275 a year.

Essay Questions

1. Explain why the classification system that categorizes countries as First World, Second World, and Third World is not as useful today as it was during the Cold War, and how the world systems theory approach has replaced it.

2. Use the concepts of the *commodity chain* and the new *international division of labor* to explain how clothing is produced in today's global economy.

3. Discuss how population characteristics are related to poverty levels around the world, specifically identifying how rich and poor nations differ on each relevant characteristic.

4. Discuss the causes of world poverty.

Solutions

PRACTICE TEST

Multiple Choice Questions

1. D, 215 One of the most common ways to measure the wealth of nations is by using the annual per capita gross national income (per capita GNI), which measures the total volume of goods and services produced per year.

2. A, 216-7 Most of the wealthy countries are in Western Europe. The poorest countries in the world are in Africa, Asia, and Eastern Europe.

3. B, 218 First world countries consist of the industrialized, capitalist nations such as the U.S., Japan, Australia, and the countries of Western Europe.

4. C, 218 Peripheral countries are poor, largely agricultural, and semi-industrialized.

5. D, 218 Semiperipheral countries such as Spain, Turkey, and Mexico are semi-industrialized. They play a middleman role in the world economy, extracting profits from poor countries.

6. D, 218 Core countries, such as the United States, control and profit the most from the world system. They rank highest on the power dimension of global stratification.

7. D, 219 The new international division of labor is characterized by increased racial inequality, the exodus of unskilled workers from poor nations into industrialized nations, and the exploitation of cheap labor in non-Western countries.

8. D, 226 According to Figure 8.2, the U.S. uses 23% of global energy, followed by China (11%), the Russian Federation (6%), India and Japan (both 5%).

9. A, 229 There has been a decline in poverty in East Asia, mostly attributable to reduced poverty in the People's Republic of China, with increases in Latin America and Southeast Asia. Sub-Saharan Africa has the highest poverty rate in the world.

10. A, 226 The gender development index measures women's well-being using three dimensions of deprivation: life expectancy, educational attainment, and income.

11. C, 230-1 Modernization theory emphasizes the role of technology and people's value systems in fostering the economic development of a country.

12. A, 221 Dependency theory explains the poverty of low-income countries as a result of their exploitation by the rich, powerful countries.

13. D, 228 The four dimensions of the human poverty index are knowledge, life expectancy, social inclusion, and economic well being.

14. C, 228 The United Nations identifies the situation in which individuals live on less than $275 per year as extreme poverty. Absolute poverty is the situation in which individuals live on less than $365 per year. The disproportionate burden of poverty carried by women in countries around the world is double deprivation.

15. A, 231 World poverty is caused by unstable or ineffective governments, collapsed economies resulting in international loans and economic restructuring imposed by the World Bank and the IMF, and changes in the world economic system. Rapid population growth does not necessarily result in poverty. Finally, world poverty is not caused by lazy people uninterested in work; people in extreme poverty work tremendously hard just to survive.

16. B, 228 Relative poverty is the amount of income that a family in the United States needs to maintain a suitable standard of living.

17. D, 232 South Korea, Malaysia, Thailand, Taiwan, and Singapore, which have shown rapid economic growth and emerged as developed nations, are referred to as the newly industrializing countries.

18. C, 229-30 Women constitute 60 percent of the world's population, perform two-thirds of work hours, receive only one-tenth of the world's income, and own only 1% of the wealth. Women are more likely than men to suffer from malnutrition, and experience greater illness and death associated with pregnancy, birth, and childrearing.

19. A, 224 The richest countries in the world have about 15 percent of the world's population, and population growth occurs at a slower rate in rich countries than in poor countries due to declining fertility rates.

20. C, 220 Anthony Marx's comparison of racial categories in Brazil and the United States indicates that being clearly labeled as Black has supported the development of group solidarity and collective action among people of color in the U.S. In Brazil, where elites declared the country a racial democracy, skin color still influences individual opportunities and social status, but the lack of clearly defined racial categories has denied Afro-Brazilians an important source of collective identity.

True-False Questions

1. T, 229 Among the industrialized nations, only Russia has a higher poverty rate than the United States [Figure 8.3].

2. T, 226 In high-income countries, education is nearly universal, but in poor countries, school attendance is low and most basic education occurs in informal settings such as the family and religious congregations.

3. T, 229 Twenty-eight percent of the world's population live in absolute poverty.

4. F, 221 Dependency theory is derived from conflict theory and views poverty as the result of neocolonialism and the expansion of capitalism.

5. F, 221 Modernization theory argues that poverty is the result of having poorly developed institutions, low technological development, and unmotivated populations who do not save or invest money.

6. T, 221-2 Dependency theory argues that core countries are able to exert greater control over the world economic system, particularly by lending money to peripheral countries, because debt perpetuates their dependence on the rich countries.

7. T, 216-7 Those countries that suffer a poor standard of living, have low levels of education and high death rates, and are generally at the bottom of the global stratification system are comprised largely of people of color.

8. F, 226 The overproduction of "greenhouse gas" emissions occurs in the United States, Canada, Australia, and parts of Western Europe and Russia.

9. T, 225 Young children are highly susceptible to waterborne illnesses such as cholera, which may result in death, especially in countries with poor access to clean water and adequate sanitation.

10. T, 225 Many of the richest countries are experiencing population declines, creating a shortage of young people to meet the society's labor needs, which may be addressed by importing workers from other countries.

Fill-in-the-Blank Questions

1. neocolonialism, 222
2. multinational corporations, 222
3. commodity chain, 223
4. world cities, 223
5. extreme, 228

Essay Questions

1. World Systems Theory. The world systems theory and the three world categorization are discussed on p. 218-219 in the text. Student responses should demonstrate an understanding of the political and economic realities on which each classification system is based.

2. Commodity Chains. Commodity chains are discussed on p. 223, and the new international division of labor is discussed on p. 219. Student responses should demonstrate an understanding of the global division of labor and the relations among countries involved in the commodity chains.

3. Rich and Poor Nations. The characteristics of rich and poor nations are discussed on p. 228-229 in the text. Student responses should demonstrate knowledge of the characteristics of both rich and poor nations and the relations between them.

4. World Poverty. The causes of world poverty are discussed in the text on p. 231-232. Student responses should demonstrate a grasp of the structural and global causes of poverty.

RACE AND ETHNICITY

BRIEF CHAPTER OUTLINE

Race and Ethnicity

    Ethnicity

    Race

    Minority and Dominant Groups

Racial Stereotypes

    Stereotypes and Salience

    The Interplay Among Race, Gender, and Class Stereotypes

Prejudice, Discrimination, and Racism

    Prejudice

    Discrimination

    Racism

Theories of Prejudice and Racism

    Psychological Theories of Prejudice

    Sociological Theories of Prejudice and Racism

Diverse Groups, Diverse Histories

    Native Americans

    African Americans

    Latinos

    Asian Americans

    Middle Easterners

    White Ethnic Groups

Patterns of Racial and Ethnic Relations

    Assimilation Versus Pluralism

CHAPTER FOCUS

This chapter explores race and ethnicity and intergroup relations, including prejudice, discrimination, stereotypes, and the forms of racism, as well as the theories explaining prejudice and racism. It also examines the diverse experiences and histories of racial and ethnic groups, and the pattern of racial and ethnic relations in the United States, including social movements and public policy debates, such as those over affirmative action.

QUESTIONS TO GUIDE YOUR READING

1. How are the concepts of ethnicity, race, minority and dominant group socially constructed?

2. What are the salience principle and stereotype interchangeability, and how do they operate in the interplay among race, gender, and class stereotypes?

3. What are the dynamics of prejudice and ethnocentrism, and how are both promoted through socialization?

4. What is racism, what are the forms of racism, and how is institutional racism embedded in the normal operation of social institutions?

5. How do the major psychological theories (scapegoat and authoritarian personality) and the major sociological theories (contact theory and the intersection perspective) explain prejudice and racism?

6. What are the diverse experiences and histories of racial and ethnic groups in the United States?

7. How do assimilation, pluralism, segregation, and the urban underclass perspectives explain the patterns of racial and ethnic relations in the United States?

8. What evidence is brought to bear in the debate about the relative importance of race and class?

9. What are the patterns of racial and ethnic conflict worldwide and the major approaches to address these problems?

10. What are the relative merits and limitations of race-specific and color-blind programs for change, as well as the achievements of and debates over affirmative action?

KEY TERMS

(defined at page number shown and in glossary)

affirmative action 262

assimilation 247

aversive racism 245

contact theory 247

discrimination 244

ethnic group 238

forms of racism 245

laissez-faire racism 245

old-fashioned racism 245

race 238

racialization 239

residential segregation 245

scapegoat theory 246

stereotype 242

urban underclass 256

anti-Semitism 255

authoritarian personality 246

color-blind racism 245

cultural pluralism 256

dominant group 241

ethnocentrism 244

institutional racism 245

minority group 241

prejudice 243

racial formation 241

racism 245

salience principle 242

segregation 256

stereotype interchangeability 243

CHAPTER OUTLINE

I. **RACE AND ETHNICITY**

    A. <u>Ethnicity</u>

       1. An **ethnic group** is a social category of people who share common cultural characteristics such as language, religion, and customs.

       2. Ethnic groups develop because of their unique historical and social experiences, which become the basis for the group's *ethnic identity*, or the definition the group has of itself as sharing a common cultural bond.

       3. Ethnic groups can develop a more or less intense ethnic identity at different points in time. For example, prejudice, hostility, and exclusionary practices from other groups often strengthen ethnic identity.

    B. <u>Race</u>

       Like ethnicity, **race** is a socially constructed category. Societies assign people to races not by logic and fact, but based on opinion and social experiences, making the definition of race a *social* process.

       1. The categories used to presumably divide groups into races are not fixed; they vary from society to society and over time. For example, the U.S. Bureau of the Census has classified various groups differently over time [Table 9.1].

2. The biological characteristics that have been used to define different racial groups vary both within and between groups. For example, in Brazil, only those people of African descent are classified as Black, regardless of skin color. On the other hand, a light-skinned Black could be considered White if the person is of high socioeconomic class; social class is thus *racialized*.

3. The biological differences presumed to define different racial groups are rather arbitrary, for example, skin color rather than hair or eye color. In fact, most of the variability in truly biological characteristics, such as blood type, is *within* (rather than between) racial groups.

4. Different groups use different criteria to define racial groups. For example, some Native American tribes recognize as members only those people who are of 75 percent ancestry, while other tribes recognize 50 percent ancestry.

5. **Racialization** is a process whereby some social category, such as social class or nationality, takes on what are perceived in the society to be race characteristics. For example, Adolph Hitler labeled Jews, an *ethnic group*, as a race, thus *racializing* them.

6. A **race** is a group treated as distinct in society on the basis of certain characteristics, some of which are biological, that have been assigned social importance; thus, race is *socially constructed*.

7. **Racial formation** is the process by which a group comes to be defined as a race through support from official institutions such as the law and schools. For example, Latinos, Asian Americans, and American Indians are defined as races in the United States despite the varied backgrounds and experiences of specific groups within each established category.

C. Minority and Dominant Groups
   1. A **minority group** is any distinct group in society that shares common group characteristics and is forced to occupy low status in society because of prejudice and discrimination.
      a. Not all racial or ethnic groups are minorities. For example, although Irish Americans were once minorities, they are not a minority group today.
      b. Minority status does not necessarily depend on numerical representation; for example, Black people in South Africa under the *apartheid* system were a numerical majority but a social minority.
   2. The group that assigns a racial or ethnic group to subordinate status in society is the **dominant group** by virtue of their power to establish such designations.
   3. A racial or ethnic group typically has four primary features:
      a. It possesses characteristics that are popularly regarded as different from those of the dominant group.
      b. It suffers prejudice and discrimination by the dominant group.
      c. Membership is frequently ascribed rather than achieved, although either form of status can be the basis for being identified as a minority.
      d. Members feel a strong sense of solidarity or "we-feeling."

II. **RACIAL STEREOTYPES**

A. Stereotypes and Salience

1. A **stereotype** is an oversimplified set of beliefs about members of a social group or social stratum that is used to categorize individuals of that group. They are presumed to describe the "typical" member of the group.

2. The categorization of people into groups and the subsequent application of stereotypes are based on the **salience principle**. Salience implies that we categorize people on the basis of what initially appears prominent and obvious about them, such as skin color, gender, and age.

B. The Interplay Among Race, Gender and Class Stereotypes

1. *Gender stereotypes* about women are more likely to be negative than those about men. The mass media convey and support cultural stereotypes of women as subservient and overly emotional, and of men as insensitive and macho.

2. *Social class stereotypes* are based on assumptions about one's social position. These include the characterizations of upper class people as snooty and phony and of lower income people as dirty, lazy, and violent.

3. The principle of **stereotype interchangeability** holds that stereotypes, especially negative ones, are often interchangeable from one targeted group to another. For example, ethnic jokes often interchange groups as the butt of the humor, but stereotype the groups in the same ways; for example, as lazy and inept.

4. Whatever group occupies low social status at a given time is stereotyped, and the stereotype is used to explain observed behavior and justify the low social status.

III. **PREJUDICE, DISCRIMINATION, AND RACE**

A. Prejudice

1. **Prejudice** is the evaluation of a social group, and individuals within that group, based on conceptions about the social group that hold together despite facts that contradict it.

2. Prejudice involves both prejudgment and misjudgment. A prejudiced person will have negative attitudes about members of an *out-group*, or any group other than one's own, as well as positive attitudes about members of one's own *ingroup*.

3. Prejudice based on race or ethnicity is called *racial-ethnic prejudice. Gender prejudice* is a negative evaluation of someone based on gender, and *class prejudice* is a negative evaluation of someone solely on the basis of social class.

4. Prejudice is revealed in **ethnocentrism**, or the belief that one's group is superior to all other groups, and has a marked effect on people's political views and voting behavior.

5. *Prejudice and Socialization.* Stereotypes and prejudices are learned and internalized through the socialization process, by all agents of primary and secondary socialization, and especially by the media.

B. Discrimination

Discrimination is the negative and unequal treatment of the members of some social group solely because of their membership in that group. *Racial-ethnic discrimination* is the unequal treatment of a person on the basis of race or ethnicity. Discrimination toward Blacks in housing, for example, is especially prominent in the United States. Pervasive discrimination in housing has resulted in persistent **residential segregation**, or the spatial separation of racial and ethnic groups into different neighborhoods.

C. Racism
   1. **Racism** is the perception and treatment of a racial or ethnic group, or member of that group, as intellectually, socially, and culturally inferior to one's own group. Racism may be overt, subtle, or covert.
   2. **Forms of racism** include: **old-fashioned** or *traditional* **racism** (or *Jim Crow racism*), **aversive racism**, **laissez-faire** or *symbolic* **racism**, and **color-blind racism**.
   3. **Institutional racism** is negative treatment and oppression of one racial or ethnic group by society's existing institutions based on the presumed inferiority of the oppressed group.
      a. Institutional racism occurs when dominant groups have the economic and political power to subjugate the minority group, even if they do not have the explicit intention of being prejudiced or discriminatory.
      b. Institutional racism is evident in the American criminal justice system in the practice of *racial profiling*, whereby police use the criterion of race to determine suspicion of criminal activity.
   4. *Institutional racism* exists at the level of social structure and *can exist even without prejudice being the cause.*

IV. **THEORIES OF PREJUDICE AND RACISM**

   A. Psychological Theories of Prejudice
      1. **Scapegoat theory** argues that historically, members of the dominant group in the United States have harbored various frustrations in their desire to achieve social and economic success. This frustration results in anger and aggression being directed toward minority groups.
      2. Another social psychological theory focuses on the personality traits of prejudiced individuals. The **authoritarian personality** is characterized by a tendency to rigidly categorize other people, submit to authority, rigidly conform, be very intolerant of ambiguity, and be inclined to superstition. Individuals with an authoritarian personality are more likely to be prejudiced.

   B. Sociological Theories of Prejudice and Racism
      1. *Functionalist theory*. Functionalist theory argues that social stability requires the assimilation of racial-ethnic minorities and women.
         a. **Assimilation** is a process by which a minority becomes socially, economically, and culturally absorbed within the dominant society.
         b. How rapidly a group assimilates into a society will depend partly on its unique history and the group members' desire to assimilate.
      2. *Symbolic interaction theory*. Symbolic interaction theory examines the role of social interaction in reducing racial-ethnic hostility and the social construction of race and ethnicity. **Contact theory** argues that interaction between Whites and minorities will reduce prejudice on the part of both groups if three conditions are met.
         a. The contact must be between individuals of *equal status*.
         b. The contact between equals must be sustained over time.
         c. Participants must agree on social norms favoring equality.
      3. *Conflict theory*. Conflict theory assumes that class-based conflict is an inherent part of social interaction; therefore, class inequality must be reduced to lessen racial and ethnic conflict

in society. The theory also focuses on the interaction of class, race, and gender through the *intersection perspective*.

V. DIVERSE GROUPS, DIVERSE HISTORIES

The histories of various racial and ethnic groups in the United States are similar in some ways, yet unique in others. The groups' histories are related by the common experience of White supremacy, economic exploitation, and political disenfranchisement. Members of some White ethnic groups, such as Irish Americans have also been victims of prejudice and discrimination.

A. Native Americans
   1. The size of the indigenous population in North America in 1492 has been estimated from one to ten million people. At the time of the first European contacts with Native Americans in the 1640s, there was considerable linguistic, religious, governmental, and economic heterogeneity among the original American Indian nations. Much of this tribal culture has since been destroyed.
   2. Government policies forced many Native Americans into inhospitable country, leading to starvation. Massive numbers of indigenous people were also killed by disease and wars of extermination, resulting in a population of only 300,000 by 1850.
   3. Today, about 55 percent of Native Americans live on or near a reservation, a system associated with abject poverty, very high unemployment, and severely limited access to education, health care, and other basic services.

B. African Americans
   1. Slaves were forcibly imported from Africa to provide free labor for sugar and tobacco plantations in the United States. Between 20 and 100 million Africans were transported to the Americas, with the vast majority going to Brazil and the Caribbean, and only 6 percent to the United States.
   2. Slavery, in which people were *chattel*, evolved as a rigid *caste system*, based on patriarchy and White supremacy. Research has revealed extensive slave resistance and armed rebellion.
   3. After Slavery presumably ended by the Civil War, the system of sharecropping emerged as a new exploitive system.
   4. The Great Migration of Black from the South to the urban North from the 1900s through the 1920s and beyond led to the formation of ghettos that subjected Blacks to grim urban conditions, but also encouraged the development of collective political, social, and cultural action.

C. Latinos

Latino or *Hispanic* Americans include Chicanos and Chicanas (Mexican Americans), Puerto Ricans, Cubans, and other recent Latin American immigrants to the United States, as well as Latin Americans who have lived in the United States for generations. There is great structural and cultural diversity within the Hispanic population, which has grown considerably in recent years.

   1. *Mexican Americans*. Chicanos lost claims to huge land areas (which became Texas, New Mexico, and parts of other Midwestern states) in the Mexican-American War of 1846-1848.

a. American immigration policy in the 1920s disproportionately restricted Mexican immigration.

b. In the early twentieth century, irrigation and its resulting year-round crop production increased the need for field labor, leading to the exploitation of migrant workers from Mexico as a cheap source of labor.

2. ***Puerto Ricans***. The Jones Act extended United States citizenship to Puerto Ricans in 1917, and the Commonwealth of Puerto Rico was established in 1952 with its own constitution. In the 1960s and 1970s, unemployment in Puerto Rico became so severe that in the U.S. government attempted to reduce the population by encouraging forms of population control, including female sterilization.

3. ***Cubans***. Cuban migration to the United States is recent in comparison to other Hispanic groups. Many of the immigrants who arrived shortly after the 1959 revolution led by Fidel Castro were middle and upper class landowners and professionals. The most recent wave of Cuban immigration occurred in 1980 when the Cuban government opened the port of Mariel to anyone who wanted to leave. The second group of immigrants has not been unable to achieve much social mobility.

D. <u>Asian Americans</u>

Asian Americans hail from many countries and have diverse cultural backgrounds.

1. ***Chinese***. Chinese Americans began migrating to the United States during the mid-nineteenth century in response to a demand for labor. They performed much of the most difficult and dangerous work of building the Central Pacific Railroad. Ethnic antagonisms led to the establishment of several urban Chinatowns on the West Coast, which were ethnic enclaves that provided support to residents.

2. ***Japanese***. The first generation of Japanese immigrants, who arrived between 1890 and 1924, were generally employed in agriculture or small Japanese businesses. The second generation of Japanese Americans became better educated than their parents, lost their Japanese accents, and generally assimilated.

a. By executive order of President Roosevelt, much of the West Coast Japanese American population had their assets frozen and their real estate confiscated by the government, and were forced to move into relocation centers during World War II.

b. In 1987, the United States government offered an official apology for their actions and awarded $20,000 to each former detainee.

3. ***Filipinos***. In 1934, the Philippine Islands became a commonwealth of the United States and immigration quotas were imposed upon the Filipinos. Over 200,000 Filipinos immigrated to the United States between 1966 and 1980, most of who were professional workers with high levels of education.

4. ***Koreans***. Korean Americans, who are largely concentrated in Los Angeles, California, tend to be former professionals who experience downward social mobility when they migrate to the United States. Conflict between African American residents and Korean business owners exists in many urban communities today.

5. *Vietnamese*. Vietnamese people began to arrive in the United States after the fall of South Vietnam in 1975, and a second wave of Vietnamese arrived after China attacked Vietnam in 1978. Despite initial discrimination, most Vietnamese heads of households in the United States are now employed full-time.

E. Middle Easterners

Immigrants from countries such as Syria, Lebanon, Egypt, and Iran speak no single language and follow no single religion, yet they are grouped together socially. Like other immigrants, many have downward mobility once they immigrate to the United States and have formed their own ethnic enclaves. Many Middle Easterners were harassed, attacked, racially profiled, and came under suspicion after the September 11th attacks.

F. White Ethnic Groups
    1. White Anglo-Saxon Protestants (WASPs) who immigrated from England, Scotland, and Wales were the first ethnic group to have widespread contact with Native American Indians. WASPs dominated the newly emerging society earlier than any other White ethnic group.
        a. The original WASP immigrants were skilled workers with a strong Protestant Ethic, or desire to work and achieve wealth.
        b. WASPs began to direct prejudice and discrimination toward other European immigrants during the mid- to late-nineteenth century. The dominance of WASPs in the U.S. has declined somewhat since 1960.
    2. There were two waves of immigration of White Ethnic groups in the 19th century, with immigrants from Northern and Western Europe arriving in the United States from 1850 to 1880 and immigrants from Eastern and Southern Europe arriving from 1890 to 1914. The Irish arrived in large numbers in the mid-nineteenth century as a consequence of food shortages and massive starvation in Ireland.
    3. Another large immigrant group was Jewish people. Over 40 percent of the world's Jewish population now lives in the United States, largely as a result of fleeing European **anti-Semitism**.
    4. In 1924, the National Origins Act imposed *ethnic quotas* that permitted immigrants to enter only in proportion to their numbers already in the country.

VI. **PATTERNS OF RACIAL AND ETHNIC RELATIONS**

A. Assimilation Versus Pluralism
    1. The *assimilation perspective* asserts that to overcome adversity and oppression, minorities need only to imitate the dominant White culture.
    2. Although many Asian American groups have followed this pattern and been identified as the "model minority," they are still subject to prejudice and discrimination, and some Asian American groups have high rates of poverty.
    3. This model is problematic because it does not consider the amount of time that it takes for certain groups to assimilate. For example, groups from rural areas typically take longer to assimilate than do groups with urban backgrounds.
    4. The assimilation model does not take into account that Blacks arrived in the United States involuntarily and were enslaved, so their histories cannot be compared to Whites who voluntarily immigrated.

5. Many White ethnics entered the United States at a time when the economy was rapidly growing and labor was in high demand, thereby providing more opportunities for attaining education and job skills despite discrimination.

6. The assimilation model raises the question of whether a society can maintain **cultural pluralism**, defined as different groups in society maintaining their distinctive cultures while coexisting peacefully with the dominant group. The Amish in Pennsylvania are an example of successful cultural pluralism.

B. Segregation and the Urban Underclass

1. **Segregation** refers to the spatial and social separation of racial and ethnic groups.

    a. *De jure segregation*, or legal segregation, is prohibited by law today.

    b. *De facto segregation*, or "in fact" segregation, still exists, particularly in housing and education.

2. According to Wilson, segregation has contributed to the creation of an **urban underclass**, or a grouping of people, largely minority and poor, who live at the absolute bottom of the socioeconomic ladder in urban areas. The problems of the inner city, such as joblessness and crime, arise from inequalities in the social structure and have dire consequences at the individual level, including teen pregnancy and AIDS.

C. The Relative Importance of Class and Race

In *The Declining Significance of Race* (1978), Wilson contended that the importance of social class has increased as the significance of race in determining Black people's access to privilege and power in the United States has declined. More recently, Wilson argued that both class and race combine to oppress not only many urban Blacks, but Whites and Hispanics, as well. Critics of his argument maintain that race is still critically important, which can be seen when one compares Blacks of a given social class to Whites of the same class.

VII. ATTAINING RACIAL AND ETHNIC EQUALITY: THE CHALLENGE

Throughout the world, conflicts stemming from racial and ethnic differences are frequently the basis for economic inequality, cultural conflict, and war.

A. Civil Rights

The major force behind most progressive social change in race relations was the civil rights movement of the 1950s and 1960s. Based on the passive resistance philosophy of Dr. Martin Luther King, Jr., learned from the philosophy of satyagraha of Mahatma Gandhi, the movement encouraged resistance to segregation through nonviolence. The movement culminated in the passage of the Civil Rights Bill and the Voting Rights and Fair Housing Acts of the 1960s.

B. Radical Social Change

1. By the late 1960s, militant leaders had grown increasingly dissatisfied with the limitations and slow process of the civil rights agenda, and the militant Black Power movement emerged.

    a. Political activist Stokely Carmichael saw inequality stemming from the institutional power that Whites had over Blacks

    b. Prior to his assassination in 1965, Malcolm X advocated a form of pluralism by demanding separate business establishments, banks, churches, and schools for Black Americans.

c. When militant groups such as the Black Panthers advocated fighting oppression with armed revolution, the United States government responded by imprisoning and killing some group members.

2. The Black Power movement dramatically altered the nature of political struggle and race and ethnic relations in the United States and influenced the development of other groups concerned with institutional racism, including the Chicano organization, La Raza Unida, and the American Indian Movement (AIM).

3. Research by the Harvard University's Civil Rights Project indicates that schools are now *more segregated* than they were 30 years ago.

C. Affirmative Action

1. *Color-blind policies* advocate that all groups be treated alike, with no barriers to oppression posed by race, gender, or other group differences.

2. *Race-specific policies* recognize that certain racial groups occupy a unique status because of a long history of discrimination and the continuing influence of institutional racism.

3. **Affirmative action**, a heavily contested program for social change, is a race-specific policy for reducing job and educational inequality. It includes two components:

    a. recruiting minorities from a wide base in order to ensure consideration of groups that have been traditionally overlooked, but without rigid quotas based on race or ethnicity; and

    b. using admission slots (in education) or designated contracts or jobs (in employment) to assure minority representation.

4. Legal decisions on the state and federal level continue to challenge affirmative action and related strategies, while the Legal Defense Fund (LDF) of the NAACP has argued forcefully for affirmative action (although not rigid quotas).

5. In 2003, the U.S. Supreme Court decided 2 cases modifying the 1978 decision that race could be used as a criterion for admission to higher education or for job recruitment as long as rigid quotas were not used. The decisions ruled out the use of a point system interpreted as a type of quota, but allowed race to be used as a factor in admissions decisions, along with other factors.

INTERNET EXERCISES

1. The *Poverty and Race Research Action Council* (www.prrac.org/) focuses on the intersection of poverty and race. Have students select one of several issues, from race/racism to homelessness and community organizing. They will find articles and other resources for extensive research.

2. *American Indian History and Related Issues* (www.csulb.edu/projects/ais) is developed and maintained by Professor Troy Johnson of California State University-Long Beach. You will find extensive links to sites on American Indian history and culture, including Indians of North and Central America and Mexico, tribe and nation homepages, American Indian Colleges, and related websites. Browse the site and select an entry that interests you and share it in class.

INFOTRAC EXERCISES

Conduct a keyword search using InfoTrac College Edition to extend the discussion in DOING SOCIOLOGICAL RESEARCH: *American Apartheid* (p. 258).

1. **Keyword: Residential segregation.** This search yields article on its prevalence and dynamics, and links to urban racial inequality, health disparities, mortality, and assisted housing.

2. **Keyword: Institutional racism.** This search leads to articles from Britain and the U.S. on healthcare, skills and preferences as factors in employment, education, judicial conduct and law enforcement.

3. **Keyword: Racial segregation.** The search leads to articles on schools, housing, health, and socioeconomic status.

4. **Keyword: Hypersegregation.** The search yields a few articles on residential segregation, using terms like spatial concentration and caste.

PRACTICE TEST

Multiple Choice Questions

1. The process whereby some social category such as nationality or religion takes on what are perceived in the society to be race characteristics is referred to as _____.
   a. discrimination
   b. stratification
   c. racialization
   d. assimilation

2. Sociologists refer to a social category of people who share common cultural element like language, religion, norms, and customs, such as Irish Americans or African Americans, as a(n) _____.
   a. dominant group
   b. racial minority
   c. ethnic group
   d. underclass

3. According to sociologists, which of the following statements about race is (are) **true**?
   a. Race is a fixed, biologically-based category that varies little over time.
   b. Definitions of race do not vary significantly within or across groups because they are based on a universal set of physical characteristics.
   c. A race is a group of people who have been defined as distinct in society on the basis of certain identifiable characteristics.
   d. all of these choices

4. Which immigrant group was initially tolerated because they provided cheap labor for the expanding railroad system, but later became the targets of prejudice and discrimination because Whites began to view them as competing for scarce jobs on the West coast in the 1800s?
    a. Cuban
    b. Chinese
    c. Africans
    d. Filipinos

5. Sociologists refer to any socially distinct group that presumably shares common characteristics and is forced to occupy low status in society because of prejudice and discrimination as a _____.
    a. dominant group
    b. racial minority
    c. minority group
    d. ethnic enclave

6. The basic premise of contact theory is that _____.
    a. individuals who have particular personality traits, such as tendencies toward rigidly categorizing other people, submitting to authority, and being intolerant of ambiguity, are more likely to be prejudiced
    b. WASPs deserve to receive a greater proportion of society's rewards because they have been living and working in the United States longer than any other group
    c. prejudice between Whites and Blacks can be reduced through providing ample opportunities for sustained interaction between group members of equal status
    d. the urban underclass developed as a result of economic and political policies designed to reduce public services (such as police patrols) in urban areas as a way to save money

7. As defined by sociologists, minority groups typically have which of the following characteristics?
    a. Members feel a strong sense of solidarity
    b. Membership is usually ascribed rather than achieved
    c. Most members share an easily identifiable characteristic
    d. all of these choices

8. Sociologists refer to an oversimplified set of beliefs about members of a social group that is used to categorize individuals of that group and justify discrimination as _____.
    a. prejudice
    b. stereotype
    c. victimization
    d. discrimination

9. Which concept implies that we categorize people on the basis of what appears initially prominent and obvious about them, such as skin color, gender, and age?
    a. salience principle
    b. victim hypothesis
    c. racial formation
    d. authoritarian personality model

10. Conflict theory argues that the best solution for reducing racial-ethnic inequality is to _____.
    a. reduce the amount of contact racial and ethnic minorities have with Whites
    b. encourage minority groups to better assimilate into the dominant culture
    c. establish stricter punishments for people who insult or assault minorities
    d. reduce the degree of class inequality in society

11. Sociologists define the overt negative and unequal treatment of the members of a social group solely because of their membership in that group as _____.
    a. prejudice
    b. discrimination
    c. differentiation
    d. ethnocentrism

12. The current spatial segregation of racial and ethnic groups into different housing areas of the United States represents which type of segregation?
    a. de jure
    b. de facto
    c. quid pro quo
    d. in loco parentis

13. Which theory argues that members of the dominant group in the U.S. have historically harbored frustrations in their desire to achieve social and economic success, which they directed toward minority group members?
    a. assimilation
    b. resentment
    c. scapegoat
    d. contact

14. The largest community of Jewish people in the world lives in which nation?
    a. Israel
    b. Poland
    c. Germany
    d. United States

15. When the United States annexed the land that became Texas in the mid-1800s, which ethnic group became defined as a race and subsequently stereotyped as lazy and corrupt?
    a. Cubans
    b. Italians
    c. Mexicans
    d. Puerto Ricans

16. Which of the following statements about WASPs is (are) **true**?
    a. WASPs immigrated primarily from Ireland and Poland.
    b. The original WASP immigrants were skilled workers with a strong work ethic.
    c. The dominance of WASPs in U.S. society has increased somewhat since 1960.
    d. all of these choices

17. Which of the following is characteristic of laissez-faire racism?
    a. Subtle but persistent negative stereotyping of minorities, particularly Black Americans, especially in the media.
    b. A tendency to blame Blacks themselves for the gap between Blacks and Whites in socioeconomic standing, occupational achievement, and educational achievement.
    c. Resistance to meaningful policy efforts designed to ameliorate America's racially oppressive social conditions and practices.
    d. all of these choices

18. The 1924 National Origins Act _____.
    a. explicitly prohibited African Americans from marrying people of a different race
    b. allowed the United States to legally establish Puerto Rico as a Commonwealth
    c. allowed an increase in immigration for groups fleeing war in their own nations
    d. imposed ethnic quotas restricting immigrants to enter the country only in proportion to their current representation

19. Those policies that recognize the unique status of racial groups because of a long history of discrimination and the continuing influence of institutionalized racism are _____.
    a. color-blind
    b. race-specific
    c. domestic-based
    d. culturally-oriented

20. The majority of Cuban people who immigrated to the United States following the Cuban revolution led by Fidel Castro were _____.
    a. impoverished people who had little education and few job skills
    b. working-class people who were moderately educated, skilled craftspeople
    c. middle-class people who were highly educated professionals and land owners
    d. None of the above groups migrated to the United States after the Cuban revolution because President Kennedy feared retribution from Castro if he allowed refugees to enter.

21. By executive order of President Roosevelt, which of the following groups had their assets frozen and their real estate confiscated by the American government, and were ordered into relocation centers in the United States during World War II?
    a. Japanese Americans
    b. German Americans
    c. African Americans
    d. none of these choices

22. Which of the following statements about Native Americans in the contemporary United States is (are) **true**?
    a. Only 10 percent of the Native American population still lives on a reservation.
    b. There is more than 50 percent unemployment among Native American men.
    c. Only African Americans have a higher poverty rate than Native Americans.
    d. all of these choices

True-False Questions

1. One necessary characteristic of a minority group is that it be comprised of fewer members than are in the dominant group.

   TRUE or FALSE

2. Two Supreme Court decisions in 2003 overturned the 1978 decision, and found that affirmative action programs in higher education are illegal, and that race cannot be used as a criterion for admission under any circumstances.

   TRUE or FALSE

3. Prejudice is a negative evaluation or attitude, while discrimination is an overt behavior.

   TRUE or FALSE

4. Imposed in 1924, the National Origins Quota Act prohibited Chinese people from immigrating to the United States.

   TRUE or FALSE

5. The authoritarian personality is characterized by a tendency to challenge authority and be very tolerant of ambiguity.

   TRUE or FALSE

6. Social psychologists argue that people with an authoritarian personality are more likely to be prejudiced.

   TRUE or FALSE

7. Functionalist theory argues that the best solution for reducing racial-ethnic inequality is to encourage segregation, which reduces opportunities for adversarial groups to interact.

   TRUE or FALSE

8. Over half of the Filipino population that migrated to the United States between 1966 and 1980 were well-educated professionals.

   TRUE or FALSE

9. The Emancipation Proclamation of 1863 affected the status of Blacks in society by encouraging the development of political action groups that supported social movements for equality.

   TRUE or FALSE

10. The United States classifies immigrants from Middle Eastern countries such as Syria, Lebanon, and Iran as one cohesive ethnic group because they share a common language and religious affiliation.

    TRUE or FALSE

Fill-in-the-Blank Questions

1. Prejudice is revealed in _____ , or the belief that one's own group is superior to all other groups.

2. The principle of stereotype _____ holds that the same negative stereotypes are often applied to various groups who are the targets of racial or ethnic prejudice.

3. The social majority, or _____ group, assigns particular racial and ethnic groups to subordinate status in society.

4. Although _____ segregation is prohibited in the United States today, _____ segregation in housing and education persists.

5. Racial profiling is an example of _____ racism, which exists at the level of social structure rather than in individual attitudes or behavior.

Essay Questions

1. Compare and contrast the experiences of three minority groups in the United States based on the circumstances and timing of their immigration.

2. Provide evidence to support the statement, "Race and ethnicity are socially constructed."

3. Identify the three conditions that contact theory argues must be met to reduce prejudice between groups, and suggest one way that these conditions could be established in an educational or workplace setting.

4. Explain four problems of applying the assimilation model to the experiences of African Americans.

5. Explain affirmative action, and describe the main arguments in favor of this race-specific policy.

Solutions

PRACTICE TEST

Multiple Choice Questions

1. C, 239 The process whereby some social category such as nationality takes on what are perceived in the society to be race characteristics is racialization. An example of this process was Hitler's labeling Jews, an ethnic religious group, as a race.

2. C, 238 Sociologists designate a category of people who share common cultural elements as an ethnic group. Only those racial and ethnic groups that are assigned to a low status in society and subject to discrimination are considered minority groups.

3. C, 238-9 Sociologists view race as a socially constructed, rather than biologically-determined, category. The designation of a category of people as a "race" is a social process. Racial classifications vary cross-culturally and over time.

4. B, 253 Chinese immigrants were a source of cheap labor for the railroad system; however, they were forced out of railroad employment in the late 1800s and subsequently faced considerable prejudice and discrimination because Whites viewed them as competition for scarce jobs in the West.

5. C, 241 A minority group is forced to occupy low status in society because of prejudice and discrimination. The group granted the authority to designate other groups as minorities is referred to as the dominant group.

6. C, 247 Contact theory posits that prejudice can be reduced by facilitating regular, sustained interaction between members of different racial-ethnic groups.

7. D, 241-2 Sociologists identify four typical characteristics of minority groups: membership is ascribed rather than achieved, members possess characteristics that are popularly regarded as different from those of the dominant group, members suffer prejudice and discrimination, and members share a strong sense of solidarity or belonging.

8. B, 242-4 An oversimplified set of beliefs about members of a social group that is used to categorize individuals of that group and justify inequality is a stereotype. Stereotypes are presumed to describe "typical" members of a social group. Prejudice is the negative evaluation of a group, whereas discrimination is the negative treatment of individuals belonging to a group based solely on their membership in that group.

9. A, 242 The salience principle implies that we categorize people on the basis of what appears initially prominent and obvious about them, such as skin color and sex.

10. D, 247-8 Conflict theory views racial and ethnic inequality as rooted in class inequality. This theory advocates challenging current social arrangements that are based on differences in power and access to valued resources. Conflict theory views social change as emerging from the active resistance of minority groups.

11. B, 244 The negative and unequal treatment of the members of some social group solely because of their membership in that group is discrimination. Prejudice is an attitude based on the negative evaluation of a group. Prejudice is revealed in ethnocentrism, or the belief that one's own values, norms, and practices are superior to those of other groups.

12. B, 256 *De jure* segregation, or legal segregation, is prohibited in the United States, but *de facto* segregation persists, particularly in education and housing.

13. C, 246 Scapegoat theory asserts that members of the dominant group have historically harbored frustrations in their desire to achieve social and economic success, which they then direct toward minority group members. Contact theory, which is associated with symbolic interaction theory, focuses on reducing prejudice through fostering positive, sustained social interaction between members of different racial groups. Functionalist theory advocates the assimilation of minority groups into the dominant culture as a mechanism for fostering social equality and social stability.

14. D, 255 Forty percent of the world's Jewish population lives in the United States, making it the largest population of Jews in the world.

15. C, 251 Mexicans became defined as a distinct racial-ethnic group based on their geographic origin, which exemplifies the process known as racial formation. White Americans stereotyped them as lazy and corrupt.

16. B, 254 The original WASPs (White Anglo Saxon Protestants) were skilled workers with a strong Protestant work ethic. They immigrated primarily from England, Wales, and Scotland. The WASP influence in the United States has slightly declined.

17. D, 245 Laissez-faire racism, which came into prominence after the Second World War, involves several elements including: (1) The subtle but persistent negative stereotyping of minorities; (2) A tendency to blame Blacks themselves for socioeconomic, occupational, and educational gaps between Blacks and Whites; and (3) Clear resistance to meaningful policy efforts designed to ameliorate America's racially oppressive social conditions and practices.

18. D, 255 The National Origins Quota Act of 1924 set the first real ethnic quotas, limiting the number of new immigrants from each country to their current representation in the United States. Consequently, those from western and northern Europe were allowed to immigrate in higher numbers than those from southern and eastern Europe, who, although White, tended to be people with darker skins.

19. B, 262 Race-specific policies such as Affirmative Action recognize that racial-ethnic groups such as African Americans have a unique status because of a long history of discrimination and the continuing influence of institutional racism. Color-blind policies such as the Civil Rights Act of 1964 advocate treating members of all groups exactly the same, without regard to race or ethnicity.

20. C, 252 Cuban immigration is relatively recent compared to the immigration of other Hispanic groups. A majority of Cubans who fled the country after the revolution led by Castro were well-educated professionals and landowners. Cubans who immigrated in the 1980s have not achieved much economic or social mobility.

21. A, 253 By executive order of President Roosevelt, Japanese Americans were forced into relocation centers after the bombing of Pearl Harbor during World War II. In 1987, the United States government issued a formal apology to Japanese Americans and awarded $20,000 to each former detainee.

22. B, 249 Today, about 55 percent of Native Americans live on or near a reservation, a system that has poorly served them. Native Americans have the highest poverty rate of any group in the country, with over 50 percent unemployment for men.

True-False Questions

1. F, 241 A minority group need not be numerically smaller than the dominant group. For example, Blacks are the numerical majority in South Africa, but were a social minority under the apartheid system.

2. F, 263 The Supreme Court decided in 1978 that race could be used as a criterion for admission to undergraduate, professional, and graduate schools or for job recruitment, as long as race is combined with other criteria and as long as rigid racial quotas are not used. This decision was essentially confirmed, but modified, by the two decisions in 2003, finding that race could be used as a factor in admissions decisions, as long as race was considered along with other factors, and finding unconstitutional as a form of "quota" a "points" system that would increase the chances of minorities to be admitted.

3. T, 243-4 Prejudice is a negative evaluation, or attitude, while discrimination is a behavior.

4. F, 255 The National Origins Quota Act of 1924 set the first real ethnic quotas, limiting the number of new immigrants from each country to their current representation in the United States. The Chinese Exclusion Act of 1882 prohibited the immigration of unskilled Chinese laborers to the United States.

5. F, 246 The authoritarian personality is characterized by rigid categorization of other people, submission to authority, and intolerance of ambiguity.

6. T, 246-7 An individual with an authoritarian personality is more likely to be prejudiced.

7. F, 247 Functionalist theorists do not support segregation; in fact, they advocate that minority groups fully assimilate into the dominant culture to promote social equality, stability, and harmony.

8. T, 253 Demographers predict that Filipinos will constitute the largest group of Asian Americans within the next thirty years. Over two-thirds of those who immigrated to the United States from 1968 to 1980 were well-educated professionals.

9. F, 250-1 The Emancipation Proclamation of 1863 legally ended slavery, but the emerging sharecropping system continued exploiting Black Americans for their labor. The Great Black Migration to northern cities, which occurred from the 1900s to the 1920s, led to the development of urban ghettos such as Harlem in New York City. Although Black Americans suffered from grim urban conditions, they also developed resources such as volunteer organizations, political action groups, and cultural achievements that supported social movements for equality.

10. F, 254 Although Middle Easterners are categorized as an ethnic group in the United States, members come from a variety of countries, speak several different languages, and follow no single religion.

Fill-in-the-Blank Questions

1. ethnocentrism, 244
2. interchangeability, 243
3. dominant, 241
4. de jure, 256; de facto, 256
5. institutional, 245

Essay Questions

1. Minority Experience. The experiences of various minority groups in the United States are discussed on p. 248-255 in the text. Student responses should reflect an understanding of the circumstances and timing of their immigration and differences within minority groups.

2. Social Construction of Race and Ethnicity. The social construction of race and ethnicity is discussed on p. 238-241. Student responses should reflect an understanding of the processes of racialization and racial formation, as well as the social construction of minority group status.

3. Contact Theory. Contact theory is discussed on p. 247. Student responses should demonstrate an understanding of the three conditions contact theory posits for reducing prejudice between groups.

4. African Americans and the Assimilation Model. The assimilation model is discussed on p. 247, and African American experiences on p. 249-251. Student responses should demonstrate understanding of the assimilation model as well as African American history and experiences.

5. Affirmative Action. Affirmative action as a race-specific policy is discussed in the text on p. 262-263. Student responses should demonstrate a grasp of the principles underlying affirmative action and how it is actually implemented in the United States.

CHAPTER 10

GENDER

BRIEF CHAPTER OUTLINE

The Social Construction of Gender

Gender: Diversity Across Cultures

Sex Differences: Nature or Nurture?

Gender Socialization

The Formation of Gender Identity

Sources of Gender Socialization

The Price of Conformity

Race, Gender, and Identity

Gender Socialization and Homophobia

The Institutional Basis of Gender

Gender Stratification

Sexism and Patriarchy

Women's Worth: Still Unequal

The Devaluation of Women's Work

Balancing Work and Family

Theories of Gender

The Frameworks of Sociology

Feminist Theory

Gender in Global Perspective

Gender and Social Change

Contemporary Attitudes

Legislative Change

Chapter Summary

CHAPTER FOCUS

CHAPTER FOCUS: This chapter examines gender socialization and gender identity in the context of gender stratification. It explores the theories that explain gender inequality in work and the persistence of gender segregation, as well as the major frameworks for explaining gender inequality in general.

QUESTIONS TO GUIDE YOUR READING

1. How are sex and gender defined, and what is the relative influence of biology and culture on gender identity?

2. What are the processes of gender socialization and the formation of gender identity?

3. What is the interplay of race and gender in gender and sexual identity?

4. What is the role of sexism and patriarchy in gender stratification and in what ways can institutions be defined as "gendered"?

5. How do human capital theory, the dual labor market, gender segregation, and overt discrimination explain gender inequality in work?

6. What explains the devaluation of women's work and in what ways does it affect the work experience of men and women?

7. What are the major frameworks for explaining gender inequality?

8. What major social changes, including attitudinal and legislative changes, have affected men and women in recent years?

KEY TERMS

(defined at page number shown and in glossary)

biological determinism 269

discrimination 283

feminism 286

gender 268

gender identity 271

gender socialization 270

gendered institution 276

homophobia 275

labor force participation rate 280

matriarchy 279

occupational segregation 282

radical feminism 287

socialist feminism 286

comparable worth 290

dual labor market theory 281

feminist theory 286

gender apartheid 277

gender segregation 282

gender stratification 277

hermaphroditism 270

human capital theory 280

liberal feminism 286

CHAPTER OUTLINE

I. THE SOCIAL CONSTRUCTION OF GENDER

A. <u>Defining Sex and Gender</u>

1. **Sex** refers to one's biological identity as male or female. **Gender** is the socially learned expectations and behaviors associated with members of each sex. Sociologists view the process of becoming a man or a woman as the result of social and cultural expectations that pattern people's behavior. From the moment of birth, gender expectations influence how boys and girls are treated, with boys being given greater independence and girls being more protected by others.

2. Gender roles associated with masculinity and femininity vary greatly across cultures. For example, Western industrialized societies tend to define masculinity and femininity as opposites. In China, however, the law formally defines marriage as a relationship between equal companions who share family and household responsibilities.

3. Substantial differences in the construction of gender across social classes and racial-ethnic subcultures within a single society may also exist. For example, the Navajo Indians historically defined three possible gender roles, including the *berdache*, who were anatomically male, lived as women, and were not considered homosexual.

B. <u>Sex Differences: Nature or Nurture?</u>

1. Biology is only one component of the differences between women and men. The important question for sociologists is how biology and culture interact to produce a person's **gender identity**.

2. **Biological determinism** refers to explanations that attribute complex social phenomena to physical characteristics.

   a. Although many believe that men are more aggressive than women because of hormonal differences, reductions in testosterone levels do not predict changes in men's aggression.

   b. Although hormonal differences between males and females in childhood are minimal, boys exhibit significantly more aggression than girls.

   c. Biological explanations of inequality between men and women tend to flourish during periods of rapid social change, helping maintain the status quo by supporting claims that differences are "natural."

3. A person's sex is determined by chromosomal structure and established at conception.

   a. In addition to the presence of different genitalia, differences exist in male's and female's average length and weight at birth as well as average resting heart rate, blood pressure, and muscle mass in adulthood.

    b. **Hermaphroditism** is a condition produced when irregularities in chromosomal formation or fetal differentiation produce persons with mixed biological sex characteristics. Case studies of hermaphrodites reveal the extraordinary influence of social factors on identity.

4. *Transgendered* people are those who deviate from the binary system of gender, including transsexuals and cross-dressers. Trangendered people experience enormous pressure to fit within normative gender expectations.

II. GENDER SOCIALIZATION

Through **gender socialization**, men and women learn the expectations associated with their sex. This process affects self-concept, social and political attitudes, perceptions of others, and feelings about relationships with others. Even people who set out to challenge traditional gender expectations often find themselves yielding to the powerful influence of socialization.

A. The Formation of Gender Identity

One result of gender socialization is the formation of **gender identity**, or one's definition of oneself as female or male. Gender identity shapes how we think about ourselves and others, and influences numerous behaviors, such as the likelihood of drug and alcohol abuse, violence, depression, and aggressive driving. Studies find strong effects of gender identity on body image; adolescent girls report lower self-esteem than boys and more negativity about their body images.

B. Sources of Gender Socialization

Gender socialization is reinforced whenever gender-linked behaviors receive approval or disapproval from others.

1. *Parents* are one of the most important sources of gender socialization. For example, parents tend to assign different chores to sons and daughters. In play activities, gender norms are applied even more strictly to boys than to girls.
2. Socialization also comes from *peers*. Through play, children learn patterns of social interaction, cognitive and physical development, analytical skills, and the values and attitudes of the culture. Research indicates that boys' play is more likely to encourage violence, individualism, and hierarchy.
3. *Schools* are particularly strong influences on socialization because of the amount of time children spend in them. Teachers tend to pay more attention to boys and often express different expectations for boys and girls.
4. *Religion* is an often overlooked but significant source of gender socialization. Religious doctrines have a strong effect on the formation of gender identity. In the United States, the major Judeo-Christian religions place strong emphasis on gender differences and explicitly affirm the authority of men over women.
5. The *media* communicate strong gender stereotypes, often delivering unrealistic portrayals of women and men.
       a. Advertisements, an important outlet for the communication of gender images to the public, tend to disseminate particularly idealized, sexist, and racist images of women and men.

b. Popular culture, including greeting cards, books, songs, films, and comic strips, all contain images that represent the presumed cultural ideals of womanhood and manhood.

C. The Price of Conformity
1. A high degree of conformity to stereotypical gender expectations takes its toll on both men and women, often with negative health consequences, including eating disorders for women and steroid use for men.
2. Male socialization discourages intimacy and encourages aggression, risk-taking, independence, self-reliance, repression of emotion, and physical strength, all of which contribute to men's higher rate of injury and early death from accidents and violence.
3. Violence associated with gender roles is endemic in the United States and around the world. Sexual assault, harassment, and domestic violence are all linked to the association of gender with men's power.
4. Women who rigidly conform to a feminine gender role defined as passive and dependent experience higher rates of depression and other forms of mental illness, as well as more physical health problems, than women who adopt a variety of traits and balance multiple social roles. Eating disorders associated with distorted body image and an ideal of thinness can also put girls' and women's health at risk.

D. Race, Gender and Identity
1. Because gender identity is merged with racial identity, men and women's roles are conditioned by the social context of their experiences as members of particular racial-ethnic groups.
2. African Americans are more likely to find value in both sexes displaying a variety of traits, including assertiveness, self-reliance, and gentleness.
    a. Given their long history of paid employment, African American women's socialization emphasizes both self-sufficiency and nurturing.
    b. For African Americans, manhood is defined by self-determination, responsibility, and accountability to family and community, but power over others is not highly valued.
3. Latino men bear the stereotype of *machismo*, or exaggerated masculinity associated with sexist behavior; however, *machismo* is also associated with dignity, honor, and respect within Latino culture.

E. Gender Socialization and Homophobia

> **Homophobia** is the pervasive fear and hatred of homosexuals. This learned attitude plays an important role in gender socialization because it encourages stricter conformity to traditional expectations, especially for males, and it becomes deeply embedded in people's definitions of themselves as men and women.

F. The Institutional Basis of Gender
1. Gender is not just a matter of identity or an attribute of individuals, but also a characteristic of social institutions, where gender is a system of privilege and inequality that systematically disadvantages women relative to men.
2. **Gendered institutions** are the total pattern of gender relations, including stereotypical expectations, interpersonal relationships, and the different placement of women and men in the hierarchies of social institutions. Women who work in organizations dominated by men report

that there are subtle ways that men's importance in the organization is communicated, often resulting in women feeling like outsiders.

III. **GENDER STRATIFICATION**

Gender stratification refers to the hierarchical distribution of social and economic resources according to gender. For example, two-thirds of illiterate people worldwide are women. Gender stratification is an institutionalized system that rests on a specific belief system that supports the inequality of women and men. Although gender stratification varies cross-culturally, research indicates that women are more nearly equal to men in societies where six conditions exist:

- women's work is central to the economy
- women have access to education
- ideological or religious support for gender inequality is not strong
- men make direct contributions to household work and child care
- work is not highly segregated by sex
- women have access to formal power and authority in public decision-making

Gender stratification is multidimensional; that is, women may have freedom in some areas of life but not others. In Sweden, for example, both women and men participate in the workforce and the household, and women have a strong role in the political system, yet women's wages lag behind men's [Figure 10.3]. Japanese women are well educated and have high labor force participation, yet there are rigid gender roles within the family. Extreme gender stratification, such as the exclusion of women from public life in Afghanistan under the Taliban's rule, has been labeled **gender apartheid**.

A. <u>Sexism and Patriarchy</u>
  1. An *ideology* is a belief system that tries to explain and justify the status quo. Sexism is an ideology, but it is also a set of institutional practices and beliefs through which women are controlled because of the social significance assigned to presumed differences between the sexes.
  2. Like racism, sexism distorts reality by making behaviors seem natural when they are rooted in entrenched systems of power and privilege.
     a. An example of sexist ideology is the idea that men should be paid more than women because men are the primary breadwinners. When the idea becomes embedded in the wage structure, people need not continue to believe in the idea for it to continue to have real consequences.
     b. The common belief that that women take men's jobs away and racial-ethnic minorities advance more rapidly than Whites is a distortion of facts, because most women work in gender- and race-segregated jobs, and women are still less likely than men to get promotions and raises.
  3. **Patriarchy**, which refers to a society or group in which men have power over women, is common throughout the world. In patriarchal societies, husbands have authority over wives and men hold most of the positions of public power.
  4. **Matriarchy** exists in some societies, but not as a mirror image of patriarchy. The Minangkabau of West Sumatra have a matriarchy in which women hold economic and social power, but rule is by consensus.

B. <u>Women's Worth: Still Unequal</u>
  1. Gender stratification is especially obvious in the persistent earnings gap between women and men. The gap has closed since the 1960s, but women who work year-round and full-time still

earn an average of 76 percent of what men earn. In 2003, the median income for men was $40,668 and $30,724 for women [Figure 10.4].

2. The **labor force participation rate** is the percentage of those in a given category who are employed either part-time or full-time. By 2002, 60 percent of women and 74 percent of men were in the labor force.

3. The labor force participation rate among women has changed most dramatically among White women, because Black women were already more likely to work for pay. The employment of married women with children has tripled since 1960.

4. Changes in family patterns in contemporary society mean that more women are the sole supporters of their dependents, and a majority of women report working to support themselves or their families.

5. The Equal Pay Act of 1963 was the first federal law to require that men and women receive equal pay for equal work, yet wage discrimination persists.

6. There are four main explanations for the continued difference in wages by sex.

   a. *Human capital theory*. This theory assumes that the economic system is fair and competitive, explains gender differences in wages as the result of differences in the individual characteristics workers bring to the job. Human capital variables include age, prior experience, number of hours worked, marital status, and education level.

   b. *Dual labor market theory*. This theory contends that women and men earn different amounts because they tend to work in different segments of the labor market, reflects the devaluation of women's work, because there are usually low wages in jobs where women are most concentrated.

      i. In the *primary labor market*, jobs are stable, wages are good, benefits are likely, and opportunities for advancement exist.

      ii. In the *secondary labor market*, there is high job turnover, low wages, short or non-existent promotion ladders, few benefits, poor working conditions, arbitrary work rules, and capricious supervision. Fast food workers are in this category.

      iii. There is also an informal sector of the labor market where there is even greater wage inequality, no benefits, and little, if any, oversight of employment practices. People who are paid under-the-table to perform services for a fee are in this sector.

   c. *Gender segregation*. **Gender segregation** refers to the distribution of men and women in different jobs. They may be separated into different occupations, or different specialties within a single occupation, that are linked to wage differences.

      i. The greater the proportion of women in a given occupation, the lower the pay.

      ii. Women are concentrated in a smaller range of occupations than men; for example, two-thirds of all employed women work as either sales clerks; clerical, food service, health service, or child care workers; hairdressers; or maids.

   d. *Overt discrimination*. A fourth explanation of the gender wage gap is **overt discrimination**, or practices that single out some groups for different and unequal treatment, including *sexual harassment* and other forms of intimidation used by dominant groups to perpetuate their own advantage.

C. The Devaluation of Women's Work

1. When jobs are defined as "women's work," they become devalued and associated with less prestige and income. For example, elementary school teachers, whose job is associated with caring for children, are paid less than airplane mechanics.

2. Only a small proportion of women work in occupations traditionally thought to be men's jobs (such as soldier), and very few men work in occupations historically considered to be women's work (such as nursing). This pattern reinforces the belief that there are significant differences between the sexes.

3. When people cross the boundaries established by occupational segregation, they may be considered gender deviants. These workers feel strong pressure to assert gender-appropriate behavior, such as wearing make-up for women. Perceptions of gender appropriateness influence the likelihood of women's success at work.

D. Balancing Work and Family

Although women's participation in the workplace continues to increase, women still have primary responsibility for meeting the needs of home and family, a phenomenon Arlie Hochschild calls the "second shift." The social speedup that comes from increased hours of employment for men and women, combined with household demands, is a source of considerable stress for women in the contemporary United States.

IV. **THEORIES OF GENDER**

A. The Frameworks of Sociology
1. *Functionalist theory* argues that men and women fill complementary roles that support an arrangement that works to the benefit of society.
2. *Conflict theory* views women as disadvantaged by power inequalities between women and men that are built into the social structure.
3. Feminist scholars, drawing on symbolic interaction theory and assuming the *ethnomethodological approach*, have developed what is known as the "*doing gender*" perspective, which interprets gender as something that is accomplished through the ongoing social interactions that people have with each other. This microlevel approach does not address the structural basis of women's oppression.

B. Feminist Theory

Feminist theory refers to analyses based on **feminism** that seek to understand the position of women in society for the purpose of bringing about liberating social changes. The link between theory and action is critical to feminist theory. There are four major frameworks within feminist theory [Table 10.1].

1. **Liberal feminism** argues that inequality for women originates in past traditions that pose barriers to women's advancement. It emphasizes individual rights and equal opportunities as the basis for social justice and social reform.
2. **Socialist feminism** is a more radical perspective that views women's oppression as originating in the system of capitalism. Because capitalism exploits women as a cheap source of labor, equality for women will only come when the economic and political system is changed.
3. **Radical feminism** interprets patriarchy as the primary cause of women's oppression and asserts that the origin of women's oppression lies in men's control over women's bodies.

Radical feminists do not believe that change can come through the existing system, which is male-dominated.

4. **Multiracial feminism**, which clearly emphasizes the interactive effects of race, class, and gender in systems of domination, argues there is no single, universal experience associated with being a woman. This theory notes that different privileges and disadvantages accrue to women and men as a result of their location in a racially stratified and class-based society.

V. GENDER IN GLOBAL PERSPECTIVE

The global division of labor is acquiring a gendered component, with female workers in the poorest nations providing a cheap source of labor for manufacturing products sold in the rich, industrialized countries. Worldwide, women work as much or more than men [Table 10.2], but receive 30 to 40 percent less pay, own only one percent of all property, and are seriously underrepresented in government. There is marked inequality in the domestic sphere in Japan, while China is unusual in that there is far greater sharing of household responsibilities between men and women. Work is not the only measure of women's inferior status throughout the world. The United Nations has concluded that violence against women is a "global epidemic" that takes many forms, including domestic violence, rape, infanticide, murder, and genital mutilation. The high rates of violence against women are supported by cultural norms, women's economic and social dependence on men, and discriminatory political practices.

VI. GENDER AND SOCIAL CHANGE

The women's movement has changed how women's issues are perceived in the public consciousness, as well as generated laws that protect women's rights.

A. Contemporary Attitudes
1. Public attitudes toward gender roles have changed noticeably; for example, only a small minority of people now disapproves of women being employed while they have young children.
2. Half of all women and men now say that the ideal lifestyle is to be in a marriage in which a husband and wife share responsibilities, including work, housekeeping, and child care. Less that half of men (47%) now believe it is best for men to hold the provider role. These trends vary across social groups.
3. People's beliefs about appropriate gender roles have evolved as women's and men's lives have changed, with younger men and single men expressing more egalitarian views than older, married men.

B. Legislative Change
1. Legislation that prohibits overt discrimination against women has been in place for 40 years, including the Equal Pay Act of 1963 and the Civil Rights Act of 1964, which forbid discrimination in employment on the basis of race, color, national origin, religion, or sex.
2. Title IX of the Educational Amendments of 1972 forbid gender discrimination in any educational institution receiving federal funds, which radically altered the opportunities available to women students, particularly in athletics.
3. Passage of anti-discrimination policies does not guarantee their universal implementation. For example, male athletes still outnumber female athletes by more than two to one, and male athletes receive more scholarship support than female athletes.

4. Although a strong legal framework for gender equality exists in the workplace, equity has not yet been achieved.
    a. Because most women work in different jobs than men, the principle of equal pay for equal work does not fully address workplace inequities.
    b. Some scholars have suggested implementing **comparable worth**, or policies that pay women and men equivalent wages for jobs involving similar levels of skill. This policy creates job evaluation systems that assess the degree of similarity between different kinds of jobs.
5. Many victories in the fight for gender equity are now at risk, such as *affirmative action*. The view that there is still a need for race- and gender-conscious actions designed to address persistent injustices was upheld by the U.S. Supreme Court in the 2003 case *Gratter v. Bollinger et al*.
6. Gender inequality could be addressed by increasing the number of women in positions of public power, although feminists note that increasing women's representation in institutions will not lead to significant change without also reforming the sexism in those institutions.

INTERNET EXERCISES

1. The Feminist Majority website (www.feminist.org/) provides extensive resources for news and research on feminist issues, with several specialized departments, like women in policing, women & girls in sports, reproductive rights, and global feminism. A particularly useful tool is the monitored Feminist Internet Gateway, through which you can access sites on violence, sports, reproductive rights, work, global feminism, lesbian rights, and more. Besides using the site as a research tool, you can study the site as a window into the movement, noting how information is organized with a policy and grassroots organization goal. Many campuses have Feminist Majority chapters you may want to join, and use your experience as a service-learning project.

2. The Institute for Women's Policy Research (www.iwpr.org/) is a rich source of data on women's status as well as research and policy reports. The site has an extensive library of research reports on employment and income, politics, poverty and welfare, work and family, and health and safety that could be used for term projects. Click on the Status of Women in the States and check out your state's status and compare it to the national trends.

3. The U.S. Department of Labor: Women's Bureau website (www.dol.gov/wb) provides an excellent resource for research, especially through the National Resource and Information Center, which includes covers topics like child and elder care, equal pay, educational resources, and nontraditional occupations. Pick a topic that interests you and write a brief report on the data available on the site concerning that topic.

INFOTRAC EXERCISES

Conduct a keyword search using InfoTrac College Edition to extend the discussion in DOING SOCIOLOGICAL RESEARCH: *Eating Disorders: Gender, Race, and the Body* (p. 278-9).

1. **Keyword: Body image and race.** This search leads to articles relating body image and race to past childhood abuse, body dissatisfaction, self-esteem, perceptions of Latina, Black and White beauty, thinness and attractiveness.

2. **Keyword: Gender and body image.** Gender and body image are related to attitudes toward aging, sports participation, depression, adolescence, and contrasting male and female ideals.

3. **Keyword: Age and eating disorders.** This search leads to articles on eating disorders related to adolescence, self-concept, media exposure, psychosocial adjustment, parental influence, and school health programs.

PRACTICE TEST

Multiple Choice Questions

1. Meg Lovejoy's research on eating disorders indicates that:
   a. Black women are more likely than White women to develop eating disorders because they have such a strong desire to achieve culturally valued standards of beauty.
   b. Eating disorders are prevalent for both Black and White women, but they manifest themselves as overeating in Black women and excessive dieting in White women.
   c. Lesbian women are less likely than heterosexual women to develop eating disorders because they reject dominant cultural beauty norms in order to distinguish themselves as homosexual.
   d. The "culture of thinness" that affects the development of women's body image is now affecting many more young men, whose participation in excessive dieting has increased significantly over the past decade.

2. Research indicates that which of these nations exhibits the unusual pattern of far greater sharing of paid work and household responsibilities between men and women?
   a. United States
   b. Korea
   c. Japan
   d. China

3. Which perspective generally attributes the complex social phenomena associated with gender to differences in physical characteristics between men and women?
   a. patriarchal domination
   b. social constructionism
   c. biological determinism
   d. chromosomal geneticism

4. The first federal law to require that men and women receive equal pay for equal work was the:
   a. Affirmative Action Executive Order
   b. Civil Rights Act
   c. Equal Pay Act
   d. Title IX statute

5. Among the Navajo Indians, the berdache were considered _____.
   a. hermaphrodites who were forced to live as women because they were anatomically deformed
   b. homosexuals because they were biologically male and married other men
   c. ordinary men who adopted many female characteristics and lived as a third gender
   d. men who were born anatomically male but underwent surgery to become female because they experienced a gender identity crisis as children

6. Which of the following physical differences exists between males and females?
   a. average length and weight at birth
   b. average muscle mass and bone density in old age
   c. average heart rate and blood pressure in adulthood
   d. all of these choices

7. Which feminist framework locates the origins of women's oppression in the capitalist system and argues that the transformation of the gender division of labor will only come about with changes in the social class division of labor?
   a. liberal feminism
   b. radical feminism
   c. socialist feminism
   d. multiracial feminism

8. Which of the following statements about gender segregation in the workplace is not accurate?
   a. When men or women cross the boundaries established by occupational segregation, they are often considered to be gender deviants, may be stereotyped as homosexual, and have their "true gender identity" questioned.
   b. Men who work in occupations customarily thought of as women's work tend to experience less upward mobility than do women who enter fields traditionally reserved for men.
   c. Across the labor force, women tend to be located in those jobs that are most devalued.
   d. none of these choices

9. Research indicates that men who thoroughly internalize gender expectations and highly conform to rigid standards of masculinity are _____.
   a. more likely to die from accidents
   b. less likely to smoke and drink alcohol
   c. more likely to have closer intimate relationships with their wives
   d. all of these choices

10. The pervasive fear or hatred of homosexuals that supports rigid definitions of gender roles is called _____.
    a. hermaphroditism
    b. homoeroticism
    c. homosapiens
    d. homophobia

11. Military academies have been constructed on a strict gender order in which masculinity is built into the schools in the behaviors expected from cadets and the opportunities given to, or denied, women. This example illustrates the concept of gendered _____.
    a. roles
    b. identities
    c. institutions
    d. personalities

12. To radical feminists, the origins of women's oppression lie in _____.
    a. changes in the gendered division of labor that occurred as a result of industrialization in capitalist societies
    b. men's control over women's bodies, which is maintained through sexual and physical assault
    c. traditional socialization practices that restrict women's opportunities in the public sphere
    d. the global exploitation of women of color as a cheap source of labor

13. Which of the following groups of men are most likely to find value in women displaying a variety of "masculine" and "feminine" traits and to support women's right to work outside the home?
    a. African American men
    b. Asian American men
    c. Hispanic men
    d. White men

14. Research finds that women are more nearly equal to men in societies where _____.
    a. work is highly segregated by sex
    b. women's work is central to the economy
    c. religious support for gender inequality is strong
    d. all of these choices

15. Which of the following characteristics is not protected by the Civil Rights Act of 1964?
    a. sex
    b. age
    c. race
    d. religion

16. The set of institutionalized practices and beliefs that distort reality by making presumed differences between men and women seem natural, even though they are rooted in social systems that distribute power unequally, is _____.
    a. homophobia
    b. devaluation
    c. matriarchy
    d. sexism

17. Which perspective suggests that people "do gender" through the daily interactions they have with one another and through the interpretations they have of other's actions and appearances as consistent with "being a man" or "being a woman?"
    a. functionalist
    b. human capital
    c. dual labor market
    d. ethnomethodological

18. Which perspective explains gender differences in wages as the result of differences in the individual characteristics (such as education level) that workers bring to their jobs?
    a. symbolic interaction
    b. dual labor market
    c. human capital
    d. glass ceiling

19. Which segment of the labor market is associated with high job turnover, short or non-existent promotion ladders, few or no benefits, poor working conditions, and arbitrary work rules?
    a. secondary
    b. primary
    c. overt
    d. dual

20. The pattern in which male and female workers are systematically separated into different occupations that are usually linked to different wages is gender _____.
    a. worth
    b. ceiling
    c. segregation
    d. participation

21. The purpose of the federal legislation commonly referred to as Title IX was to:
    a. forbid discrimination against women in any educational program, including sports, at any school that receives government funding.
    b. increase the percentage of women working in traditionally male occupations by enforcing quotas that required those employers to hire more women.
    c. redress past discrimination in education by providing government funding for colleges to recruit and retain female students in science, math, and engineering.
    d. forbid discrimination in employment practices, including hiring and promotion.

22. Which theory has asserted that men fill instrumental roles and women fill expressive roles in society, thereby creating an efficient social arrangement?
    a. conflict theory
    b. feminist theory
    c. functionalist theory
    d. gendered institutions theory

23. Which of the following statements about women's labor force participation is accurate?
    a. Currently, 60% of all women are in the paid labor force, compared with 74% of men.
    b. Two-thirds of mothers are now in the labor force, including more than half of mothers with infants.
    c. Women who work year-round and full time still earn, on average, only 76% of what men earn.
    d. all of these choices

24. Which feminist framework asserts that gender is learned through traditional patterns of socialization and change can best be accomplished through legal reform?
    a. liberal feminism
    b. radical feminism
    c. socialist feminism
    d. multiracial feminism

25. The principle of paying women and men equivalent wages for different jobs that involve similar levels of skill is _____.
    a. affirmative action
    b. comparable worth
    c. doing gender
    d. glass ceiling

True-False Questions

1. Human capital variables include factors such as one's weight, height, and heart rate.

   TRUE or FALSE

2. The term transgendered refers to an individual who is sexually attracted to people of the same sex.

   TRUE or FALSE

3. According to dual labor market theory, women are more likely than men to work in the primary labor market.

   TRUE or FALSE

4. Research shows that girls are more competitive when they are playing in same-sex groups, but give in to boys' power when they play with boys.

   TRUE or FALSE

5. On average, women in the United States today earn sixty percent of what men earn.

   TRUE or FALSE

6. The major Judeo-Christian religions place strong emphasis on gender equality, with explicit support for an equal division of labor and decision making between men and women in both the public and private spheres.

   TRUE or FALSE

7. Public opinion polls indicate that men are more likely than women to find the death penalty, medical testing on animals, and buying and wearing animal fur to be morally acceptable.

   TRUE or FALSE

8. When asked to rate desirable characteristics in men and women, White men and women are more likely than Latinos to select different traits for men and women.

   TRUE or FALSE

9. The wage gap between men and women is lowest − that is, women's wages are nearly equal to men's wages − in Switzerland, a country that strongly supports egalitarian gender roles.

   TRUE or FALSE

10. Girls who choose toys defined as "masculine" or play activities associated with boys are more negatively regarded and seriously sanctioned than are boys who choose "girl's" toys or activities.

    TRUE or FALSE

Fill-in-the-Blank Questions

1. Gender _____ , a system of stratification characterized by extreme segregation and the exclusion of women from public life, was instituted in Afghanistan when the Taliban seized power in 1996.

2. Most societies have some form of gender _____ , or the hierarchical distribution of economic and social resources according to gender.

3. The term _____ refers to a society or group in which men have power over women, often in both the public and private spheres.

4. Analyses that seek to understand the position of women in society for the purpose of bringing about liberating social changes are included in the framework known as _____ theory.

5. Sexual harassment and other means of intimidation are examples of overt _____ , or practices that single out women for different and unequal treatment.

Essay Questions

1. Explain how the mass media contribute to gender socialization in the United States, giving specific examples to support your points.

2. Describe the role of homophobia in gender socialization.

3. Compare and contrast human capital theory and dual labor market theory as explanations for the persistence of pay differences between women and men.

4. Identify the various forms of violence that are committed against women around the world, and discuss what factors contribute to the "global epidemic" of violence against women.

5. Provide several examples of laws that have increased women's opportunities in the United States, and discuss the limitations of legal remedies for promoting gender equality.

Solutions

PRACTICE TEST

Multiple Choice Questions

1. B, 278-9 Although men develop eating disorders, they tend to use steroids rather than diet excessively. Both Black and White women experience problems that result in eating disorders, but Black women are more likely to overeat, while White women are more likely to diet excessively. White women are strongly affected by the "culture of thinness." Although Black women tend to reject Eurocentric standards of beauty, racism fosters health problems.

2. D, 269 The Chinese Marriage Act of 1950 states that marriage is a union of equal companions who are expected to share child care and household responsibilities. In China, both men and women work long hours in paid employment. In Japan, there is significant gender inequality in the domestic sphere. (See also p. 288).

3. C, 269 Biological determinism refers to explanations that attribute complex social phenomena to physical characteristics. Sociologists believe that gender is socially constructed because what appears to be natural is only what people have been taught is normal through socialization.

4. C, 290 The Equal Pay Act of 1963 was the first piece of federal legislation to prohibit paying men and women different wages for the same work. The Civil Rights Act of 1964 prohibited gender discrimination in employment. Title IX of the Educational Amendments explicitly forbid gender discrimination in schools.

5. C, 268 The berdache were anatomically normal men who were defined as a third gender. They married other men but were not considered homosexuals.

6. D, 270 At birth, males tend to be longer and weigh more than females. In adulthood, men have a lower resting heart rate, higher blood pressure, and higher muscle mass and density than women.

7. C, 287 Socialist feminist locates the origins of women's oppression in the capitalist system, which exploits them both as women and laborers.

8. B, 284-5 Men who work in occupations customarily thought of as women's work tend to experience more upward mobility than do women who enter fields traditionally reserved for men. Both men and women who cross boundaries established by occupational segregation are often considered gender deviants and have their gender identity questioned. Across the labor force, women tend to be located in those jobs that are most devalued.

9. A, 273-4 Rigid conformity to traditional gender roles is associated with risk-taking behavior for men, particularly involvement in activities correlated with poor health outcomes and early death. Overconformity to the masculine role encourages emotional repression and self-reliance and discourages intimacy.

10. D, 275 Homophobia is the pervasive fear and hatred of homosexuals. It encourages conformity to gender roles by acting as a mechanism of social control.

11. C, 276-7 Gendered institutions are the total pattern of gender relations – stereotypical expectations, interpersonal relationships, and the different placement of women and men in the hierarchies of social institutions, such as the military. Gender identity refers to one's definition of oneself as female or male. Gender roles are the set of social expectations for masculinity and femininity.

12. B, 287 Radical feminism locates the origins of women's oppression in men's control over women's bodies, which is reflected in violence against women [Table 10.1].

13. A, 275 African American women and men tend to value traits such as assertiveness, athleticism, and gentleness in both sexes. Black women have a long history of paid employment, and Black men are more likely to support equal rights for women than are other groups of men.

14. B, 277 In societies where women's work is central to the economy, work is not highly segregated by sex, men contribute to housework and childcare, women are well educated, and ideology supporting gender inequality is weak, there is greater gender equality.

15. B, 290 The Civil Rights Act of 1964 forbid discrimination in employment on the basis of sex, race, color, national origin, and religion, but not age.

16. D, 277-8 Sexism is an institutionalized set of practices and beliefs that define differences between women and men as "natural." An ideology is a belief system that tries to explain and justify the status quo. Homophobia is the pervasive fear and hatred of homosexuality. A matriarchy is a society in which women are assigned power. Devaluation refers to assigning lower value to work done by a particular group, resulting in lower wages and prestige being associated with that work.

17. D, 286 "Doing gender" is a feminist approach arising from the ethnomethodological perspective that interprets gender as something that is accomplished through ongoing social interaction [Table 10.1].

18. C, 280 Human capital theory explains gender differences in wages as the result of differences in the individual characteristics that workers bring to the job, such as education level and experience. The glass ceiling refers to the subtle yet decisive barrier to advancement that women find in the workplace.

19. A, 281 Dual labor market theory contends that jobs in the secondary labor market (such as cashier) have low wages, few benefits, and poor working conditions. Jobs in the primary labor market (such as management) are relatively stable and have good wages and benefits. Women are disproportionately represented in the secondary labor market.

20. C, 282 Gender segregation refers to a pattern whereby workers are systematically distributed in different occupations on the basis of gender.

21. A, 290 Title IX of the Educational Amendments of 1972 forbid gender discrimination in any educational institution receiving federal funds. Adoption of this bill radically altered the opportunities available to female students, most notably in athletic activities. The Civil Rights Act forbids employment discrimination on the basis of race, color, religion, national origin, or sex. Affirmative Action is a method for addressing past discrimination through intentional efforts to recruit and retain socially disadvantaged groups.

22. C, 285-6 Functionalist theory traditionally argued that men fill instrumental roles and women fill expressive roles, creating an efficient arrangement that benefits society. Feminist theorists have been very critical of this assumption.

23. D, 280 Sixty percent of all women were in the paid labor force, compared with 74% of men by 2000. Two-thirds of mothers, including more than half of mothers with infants, are now in the labor force. Women who work year-round and full time still earn less than men—76% on average.

24. A, 286-7 Liberal feminism is a framework that views traditional socialization practices as the main cause of women's inequality, and suggests change through the legal system. As indicated in Table 10.1, radical and socialist feminism suggest that substantial change cannot occur within the current social system because society is characterized by patriarchy and class inequality. Multiracial feminism posits that there is no universal female experience, and change will come through forming alliances with other groups.

25. B, 290 Comparable worth is the principle of paying women and men equivalent wages for jobs involving similar levels of skill. Comparable worth policy suggests creating pay scales on the basis of job conditions rather than the sex composition of the workforce. Affirmative Action redresses past discrimination through intentional efforts to recruit and retain women and racial-ethnic minorities. The glass ceiling refers to the fact that women are substantially blocked from senior management positions in the workplace. Doing gender refers to social interaction that confirms perceptions of gender roles.

True-False Questions

1. F, 280 Human capital refers to characteristics that individual workers bring to the workplace, including age, education level, and experience.

2. F, 270 People who are attracted to others of the same sex are defined as having a homosexual sexual orientation. Transgendered people are those who deviate from the binary (either/or) system of gender, including transsexuals and cross-dressers.

3. F, 281 Jobs in the primary labor market are relatively stable, have good wages and benefits, and offer opportunities for advancement. Men are more likely than women to work in the primary labor market.

4. F, 272 Research shows that girls are more cooperative when they play in same-sex groups.

5. F, 280 On average, women who work full time, year-round earn approximately 76 percent of what employed men earn in the United States.

6. F, 272 Judeo-Christian religions explicitly affirm men's authority over women, thereby supporting gender inequality in society.

7. T, 268 Among other gender differences, a 2003 Gallup Poll indicates that men are more likely than women to consider the death penalty, medical testing on animals, and buying and wearing animal fur morally acceptable.

8. T, 275 Counter to the stereotype of Hispanics as holding highly polarized views of manhood and womanhood, White men and women are more likely than Latinos to select different trains as desirable for men and women.

9. F, 280 The wage gap is lowest in Turkey, where women earn nearly the same wages as men [Figure 10.4]. In Switzerland, women on average earn 70% of what men earn.

10. F, 272 Gender norms concerning play activities are more strictly applied to boys than to girls, resulting in boys who choose more "feminine" toys being more negatively regarded and more strongly discouraged than girls who choose "masculine" toys.

Fill-in-the-Blank Questions

1. apartheid, 277
2. stratification, 277
3. patriarchy, 279
4. feminist, 286
5. discrimination, 283

Essay Questions

1. Media. Mass media, as agents of gender socialization, are discussed on p. 272-273 in the text. Student responses should demonstrate an awareness of how media reflect and promote gender expectations.

2. Homophobia. Homophobia is discussed on p. 275-276 in the text. Student responses should demonstrate an understanding of heterosexism and homophobia and how the latter plays a key role in gender socialization to appropriate gender roles and behaviors for men and women.

3. Explaining the Pay Gap. Human capital and dual labor market theories are discussed on p. 280-281 in the text. Student responses should demonstrate a grasp of the two theories and what each contributes to explaining the persistence of pay differences between women and men.

4. Violence Against Women. Violence against women, described as a "global epidemic," is discussed on p. 288 of the text. Student responses should demonstrate an understanding the social, cultural, economic, and political factors that contribute to violence against women.

5. Legislating Gender Equality. The history of legislation related to gender and gender equality are discussed on p. 290-291 in the text. Student responses should demonstrate knowledge of the laws, their achievements, and their limitations in promoting gender equality.

SEXUALITY

BRIEF CHAPTER OUTLINE

Sex and Culture

   Sex: Is It Natural?

   The Social Basis of Sexuality

Contemporary Sexual Attitudes and Behavior

   Changing Sexual Values

   Sexual Practices of the U.S. Public

Sexuality and Sociological Theory

   Sex: Functional or Conflict-Based?

   Symbolic Interaction and the Social Construction of Sexual Identity

Sex, Diversity, and Inequality

   Sexual Politics

   The Influence of Race, Class, and Gender

   A Global Perspective on Sexuality

   Understanding Gay and Lesbian Experience

Sex and Social Issues

   Birth Control

   New Reproductive Technologies

   Abortion

   Pornography

   Teen Pregnancy

   Sexual Violence

Sex and Social Change

   The Sexual Revolution: Is It Over?

Technology, Sex, and Cybersex

Commercializing Sex

Chapter Summary

CHAPTER FOCUS

This chapter examines how sexuality, sexual identity and sexual orientation are socially constructed, and how the major sociological theories analyze sexuality. Sexual politics, contested social issues related to sexuality, the sexual revolution, changes in sexual values and practices, and the impact of technology, cybersex, and the commercialization of sex are also explored.

QUESTIONS TO GUIDE YOUR READING

1. How are sexuality, sexual identity and sexual orientation socially defined and patterned?

2. What have been the major changes in sexual values and practices of the American public in recent years?

3. How do the major sociological theories analyze sexuality?

4. What is the meaning of the term sexual politics?

5. How is sexuality influenced by race, class, and gender?

6. How does the international sex trade operate?

7. How does sociology approach the contested social issues related to sexuality, including birth control, new reproductive technologies, abortion, pornography, and sexual violence?

8. How did the sexual revolution affect attitudes and behaviors, and is it over?

9. What are the impacts of technology, cybersex, and the commercialization of sex?

KEY TERMS

(defined at page number shown and in glossary)

coming out 301

heterosexism 307

queer theory 302

sexual politics 302

sexual scripts 297

eugenics 309

homophobia 305

sexual orientation 301

sexual revolution 313

social construction perspective 301

CHAPTER OUTLINE

I. **SEX AND CULTURE**

    A. <u>Sex: Is It Natural?</u>

        Sexuality is socially defined and patterned. What appears to be natural is learned, based on accepted cultural customs and sanctioned by social institutions. While there is discussion of a possible *gay gene*, the evidence is skimpy and has not controlled for environmental impact.

    B. <u>The Social Basis of Sexuality</u>

        • **Human sexual attitudes and behavior vary in different cultural contexts**, resulting in different sexual behaviors being defined as normal or deviant.

        • **Sexual attitudes and behavior change over time**, as reflected by the earlier age at which young people are having sex and the increase in the number of people who have sex before marriage.

        • **Sexual identity is learned** through the socialization process and interaction with others; for example, **sexual scripts** teach us appropriate behavior for our gender, such as norms about heterosexuality and marriage.

        • **Social institutions channel and direct human sexuality** by defining some forms of sexual expression as more legitimate than others and granting certain privileges, such as the right to marry, only to heterosexual couples.

        • **Sex is influenced by the economic institutions of society**, as evidenced in the use of sex to sell products (such as cars and personal care items) as well as the sale and purchase of sex itself.

        • **Public policies regulate sexual and reproductive behavior**, as reflected in government decisions about which reproductive technologies to endorse and whether to provide federal funding for abortion and sex education.

II. **CONTEMPORARY SEXUAL ATTITUDES AND BEHAVIOR**

    Overall, Americans are more sexually liberal and show greater tolerance for diverse sexual lifestyles and practices than in the past.

    A. <u>Changing Sexual Values</u>
        1. Sexual values are currently a mix of liberal and conservative perspectives; for example, while only 38% (versus 68% in 1969) now believe premarital sex is wrong, attitudes among teens have become more conservative.
        2. Significant gender differences are evident on many topics; for example, men (62%) are more likely than women (56%) to think that homosexual relations are morally wrong.
        3. Sexual attitudes are also shaped by race-ethnicity, age, education, and religious identification; for example, sexual liberalism is associated with higher education, youth, urban lifestyle, and political liberalism on other social issues.
    B. <u>Sexual Practices of the U.S. Public</u>

Most surveys investigate sexual attitudes because self-report data about sexual practices are often unreliable. Laumann and colleagues conducted the most comprehensive survey of sexual practices in the United States in the early 1990s. The results indicated that:
- *Young people are becoming sexually active earlier.*
- *The proportion of young people who are sexually active has increased, particularly among young women.*
- *Having only one sexual partner in one's lifetime is rare.*
- *A significant number of people have extramarital sex.*
- *A significant number of people are lesbian or gay.*
- *For those who are sexually active, sex is relatively frequent.*

III. SEXUALITY AND SOCIOLOGICAL THEORY

A. Sex: Functional or Conflict-Based?

1. Functionalist theory views sexuality in terms of how it contributes to the stability of social institutions; for example, norms that restrict sex to marriage encourage the formation of heterosexual families.

2. Conflict theorists view sexuality as part of the power relations and economic inequality in society. Because sex is linked to other forms of subordination and exploitation, rape and sexual harassment are understood to be the result of power imbalances between men and women, and the international sex trade is linked to women's poverty and low social status around the world.

B. Symbolic Interaction and the Social Construction of Sexual Identity

1. Symbolic interaction theory uses a **social construction perspective** to interpret sexual identity as learned, not inborn.
   a. Although hormonal fluctuations, sexual physiology, and genetic factors are elements in sexual desire, sociologists question the extent to which biology shapes sexual identity.
   b. Sexual identity is learned through the socialization process, develops through social experiences, and is constructed over the life course through a process of self-definition.

2. **Sexual orientation**, or how individuals experience sexual arousal and pleasure, is classified as heterosexual, homosexual, or bisexual in the United States. Although some use *sexual preference* and *sexual orientation* interchangeably, gays and lesbians have argued that *sexual preference* implies a choice, whereas *sexual orientation* implies something more deeply rooted in identity.

3. **Coming out** refers to the process of defining oneself as gay or lesbian, which is a series of events in which a person comes to see herself or himself as having a gay identity and may publicly reveal that identity to others. That a person's sexual identity may change over their lifetime indicates that identity is socially created.

4. **Queer theory** is based on the social constructionist perspective and also links the study of sexuality with the study of gender.

IV. SEX, DIVERSITY, AND INEQUALITY

A. Sexual Politics

1. **Sexual politics** refers to the link that exists between sexuality and power, both in personal relationships and social arrangements.

2. Sexual politics is reflected in the sexual exploitation of women in society, the high rates of violence against women and sexual minorities, and the privilege and power accorded to those presumed to be heterosexual.

3. Both the gay and lesbian liberation movement and the feminist movement have put sexual politics at the center of the public's attention by challenging gender role stereotyping and sexual oppression.

B. The Influence of Race, Class, and Gender
1. Sexual behavior follows gendered patterns in addition to patterns established by race and class relations.
   a. The double standard is the idea that men are expected to have a stronger sex drive than women, so women who are openly sexual may be labeled "loose."
   b. Men are socialized to see sex in terms of performance and achievement, whereas women are taught to associate sex with intimacy and affection.
   c. Certain sexual stereotypes are associated with particular racial, ethnic, and social class groups; for example, Latin men are stereotyped as "hot lovers," African American men as overly virile, Asian American women as compliant and submissive, and working-class women as "sluts."
2. Class, race, and gender hierarchies have historically been justified by claiming that people of color and women are sexually promiscuous and uncontrollable.
   a. One way that White slaveowners expressed their ownership was through the sexual abuse of Black women, who were depicted as sexual animals.
   b. Black men were stereotyped as highly sexed, lustful beasts that were a threat to White women, a stereotype that supported lynching.
   c. Sexual abuse was also part of the conquest of American Indians by Whites, and the rape of women following wars is all too common.
3. Poor women and women of color are groups most vulnerable to sexual exploitation. Women who work in the sex industry (such as prostitutes and topless dancers) may have no other option for supporting themselves. Women who sell sex are usually condemned for their behavior, which is not as true for their male clients.

C. A Global Perspective on Sexuality
1. Cross-cultural studies show that sexual norms develop within particular cultural meaning systems; for example, there is considerable cross-national variation in the degree to which men and women experience jealousy when their partners kiss, flirt, or become otherwise sexually involved with another person.
2. As the world has become more globally connected, an international sex trade has flourished, linking economic development, world poverty, tourism, and the subordinate status of women in many societies.
3. The *international sex trade* (also referred to as "sex trafficking") refers to the use of women worldwide as sex workers in an institutional context, especially in "sex capitals," where sex is a commodity used to promote tourism, cater to business and military men, and support a huge nightclub industry.

4. The international sex trade has been strongly implicated in the spread of AIDS worldwide as well as the exploitation of women in countries where women have limited economic opportunities.

D. Understanding Gay and Lesbian Experience

Feminist and gay liberation movements argue that gay and lesbian experience is merely one alternative in the broad spectrum of human sexuality.

1. **Homophobia** permeates the culture, and **heterosexism** is reinforced through institutional mechanisms.
2. Lesbians and gays are a *minority group* in U.S. society, but they have organized to promote civil rights protections with some success.

V. **SEX AND SOCIAL ISSUES**

Although most people think of sex in terms of interpersonal relationships, sexual norms are deeply intertwined with various social problems in the United States. For example, teen pregnancy, pornography, and sexual violence can generate personal troubles that have their origins in the structure of society.

A. Birth Control
1. Birth control availability is now less debated than in the past, but this form of reproductive technology is still related to the position of women in society.
   a. Men mostly define the laws and make scientific decisions about what types of birth control will be available.
   b. Women are seen as more responsible for contraception; at the same time, birth control technology breaks the link between sex and reproduction, freeing women from some traditional constraints on their behavior.
2. In 1960, the federal Food and Drug Administration approved the marketing of a new oral contraceptive, the birth control pill. The Supreme Court first defined birth control as a right, not a crime, in *Griswold v. Connecticut* in 1965. This right was not extended to unmarried people until 1972 in *Eisenstadt v. Baird*.
3. **Eugenics** sought to apply scientific principles of genetic selection to "improve" the offspring of the human race. The eugenics movement, an explicitly racist and class-based trend, emerged as the birth rate fell among the White upper and middle classes in the early twentieth century.

B. New Reproductive Technologies

Although new reproductive technologies (such as surrogacy, in vitro fertilization, cloning, and gene splicing) have increased sexual freedom, they have also raised social policy questions concerning the removal of biological parents from the reproductive process. Although the concept of reproductive choice is important to most people, choice is conditioned by the constraints of race, class, and gender inequalities in society.

C. Abortion
1. Abortion is one of the most seriously contended political issues in the United States, although the majority of the American public supports abortion rights, at least in some circumstances.

2. The right to abortion was first established in constitutional law by the *Roe v. Wade* Supreme Court decision issued in 1973, which established that the government may restrict access to abortion in different trimesters of pregnancy.
3. Data on abortion show that it occurs across social groups, although certain patterns do emerge, with young women and Black and Hispanic women more likely than older women and White women to have abortions.
4. The abortion rate has declined since 1980 from a rate of 29.3 to 21.3.
5. Attitudes toward abortion are clearly rooted in more general attitudes about sexuality, family life, gender roles, and women's right to control their bodies.

D. Pornography

There is little social consensus about the acceptability and effects of pornography. Debate about pornography often focuses on the legal definition of obscenity. Public agitation over pornography has divided people into those who think it is solidly protected by the First Amendment, those who want it strictly controlled, those who think it should be totally banned for moral reasons, and those who want it banned because it harms women.

E. Teen Pregnancy
1. The United States has the highest rate of teen pregnancy among the developed nations, although teen pregnancy has declined since 1990, along with the rate of marriage for teens that become pregnant; 78% of teen births are unplanned.
2. Teenage pregnancy correlates strongly with poverty, lower educational attainment, joblessness, and health problems. Teen mothers who raise their children alone suffer the economic consequences of raising children in female-headed households.
3. Teen pregnancy is integrally linked to the gender expectations of men and women in society.

F. Sexual Violence
1. The women's movement has been successful in identifying and raising public awareness about the problems of rape, sexual harassment, domestic violence, incest, and other forms of sexual coercion and violence.
2. Feminists argue that sexual violence is a form of power relations that is shaped by the social inequality between men and women, and note that these are forms of deviant and criminal behavior, not expressions of human sexuality.
3. Various forms of sexual coercion can be understood in the context of how social institutions shape human behavior. For example, *date rape*—forced and unwanted sexual relations by someone who knows the victim—often affects young women on college campuses, particularly in organizations (such as fraternities) that define masculinity as competitive and women as sexual prey.
4. Black, Hispanic, and poor women are most likely to be victimized by violence, although White women are as likely to be victimized by an intimate partner.

VI. SEX AND SOCIAL CHANGE

A. The Sexual Revolution: Is it Over?

The **sexual revolution** has narrowed differences in the sexual experiences of men and women, and challenged gender role stereotyping and sexual oppression.

B. Technology, Sex, and Cybersex

Contraceptives and sex on the Internet (cybersex) have both changed sexual values and practices.

C. Commercializing Sex

Sex is increasingly a commodity, and sexual relationships and values are shaped by the inequalities of race, class, and gender.

INTERNET EXERCISES

1. Visit the Alan Guttmacher Institute website (www.agi-usa.org/). This institute, dedicated to reproductive health research, policy analysis, and public education, has extensive resources of data, research reports, news bulletins, and analyses accessible through the website. Access this site for research on teen pregnancy, birth control policy, sexual behavior, and other topics.

2. Advocates for Youth (www.advocatesforyouth.org/) is a comprehensive site for all issues related to adolescent sexuality, in the U.S. and in developing countries, including sex education, dating violence, youth of color, HIV/AIDS, GLBTQ youth, pregnancy and childrearing. In addition, it hosts several additional sites targeting the entertainment industry, Latino youth, HIV prevention and HIV-positive youth, youth activists and others. Explore the reports and resources on the site and link to other sites that interest you. This is the complete resource of youth and sexuality. See also the National Organization on Adolescent Pregnancy, Parenting, and Prevention (www.noappp.org/).

INFOTRAC EXERCISES

Conduct a keyword search using InfoTrac College Edition to extend the discussion in DOING SOCIOLOGICAL RESEARCH: *Teens and Sex: Are Young People Becoming More Sexually Conservative?* (p. 300-1).

1. **Keyword: Sexual conservatism.** This search leads to articles linking sexual conservatism to politics, reproductive and sexual health, heterosexual attitudes toward bisexuals, sexual harassment, and religiosity.

2. **Keyword: Sexual liberation.** This search yields articles on menopausal freedom, the religious left, prostitution, and gay marriage.

3. **Keyword: Sexual attitudes and youth.** The search leads to articles linking youth and sexual attitudes to knowledge and risk behaviors, self-perceived needs, sexual health education, and self-image.

4. Keyword: Sexual revolution and age. This search leads to articles on young women and sex, sex and the information age, and reviews of a book on the sexual revolution in Russia.

PRACTICE TEST

Multiple Choice Questions

1. Sociologists generally agree that sexual identity _____.
   a. develops in early childhood and changes little over one's lifetime
   b. is the driving force behind all human activities
   c. is biologically determined at birth
   d. none of these choices

2. According to Vasquez, men who disdain sports, work in personal service occupations such as hairdresser, or dress in ways considered feminine are at risk for assault because they engage in _____.
   a. compulsory heterosexuality
   b. gender betrayal
   c. sexual violation
   d. homosexuality

3. The sexual revolution in the United States contributed to which of the following social changes?
   a. increase in the commercialization of sex, which uses women in demeaning ways
   b. increase in the public's acceptance of sex as a normal part of social development
   c. increase in the percentage of women who have sexual intercourse before marriage
   d. all of these choices

4. Sexual politics refers to the link between sexuality and which of the following phenomena?
   a. the status of women in society
   b. the high rates of violence against women and sexual minorities
   c. the privilege and power accorded to heterosexuals
   d. all of these choices

5. Which of the following characteristics of the Catholic Church contributed to a climate wherein sexual abuse of young boys by priests could occur and be unreported for many years?
   a. The priesthood is highly patriarchal.
   b. The Catholic Church is an organization marked by a clear hierarchy of power; and it is defined as sacred.
   c. The priesthood is imbued with a sense of secrecy in the private relationship between a parishioner and a priest in the confessional.
   d. all of these choices

6. Which group of women is most vulnerable to sexual exploitation?
   a. White, middle class, teenaged girls
   b. professional and highly educated women
   c. poor women and women of color
   d. young women raised in a sexually conservative and repressed environment

7. A cross-cultural study of sexual jealousy in seven different nations found that _____.
   a. men, but not women, in all seven nations were jealous when they saw their partners kissing, flirting, or being sexually involved with another person
   b. women, but not men, in all seven nations were jealous when they saw their partners kissing, flirting, or being sexually involved with another person
   c. both men and women in all seven nations were consistently jealous when they saw their partners kissing, flirting, or being sexually involved with another person
   d. there were significant cross-national differences in the degree of jealousy among men and women in the seven nations

8. Which of the following statements about sexual identity is accurate?
   a. Coming out as a lesbian or gay is usually the result of a single homosexual experience.
   b. A person who comes out as gay or lesbian never returns to a heterosexual identity.
   c. Discovering one's sexual identity follows a predictable sequence, and once the identity is defined, it is set for life.
   d. none of these choices

9. Young children learn what is appropriate sexual behavior for people of their gender from sexual _____.
   a. revolutions
   b. activities
   c. politics
   d. scripts

10. Which of the following conclusions did Laumann and colleagues make based on their survey of sexual practices in the United States?
    a. A large percentage of Americans have sexual intercourse before marriage.
    b. Sex is fairly infrequent among people who are sexually active.
    c. Relatively few Americans have extramarital affairs.
    d. all of these choices

11. Research on U.S. teen sexual activity in the 1990s found that _____.
    a. the percentage of sexually active teens dropped from the early 1990s
    b. rates of teen pregnancy increased and teens were having more abortions
    c. the rate of sexually transmitted diseases among teens skyrocketed
    d. all of these choices

12. Barbara Risman and Pepper Schwartz's review of various studies exploring teen sexuality in the 1990s found that _____.
    a. the number of high school boys who are virgins had increased
    b. the sexual behavior of African American girls had not changed, but the sexual activity of White and Hispanic girls had increased significantly
    c. the differences between the sexual behavior of teen boys and teen girls had increased
    d. all of these choices

13. Which of the following statements about queer theory is accurate?
    a. Queer theory underscores the fact that heterosexual or homosexual attraction is fixed in biology.
    b. Queer theory encourages people to find their true sexual identity and to adhere to it throughout the life course.
    c. Queer theory interprets the various dimensions of sexuality as thoroughly social and constructed through institutional practices.
    d. While queer theory has challenged the idea that heterosexuality is the only form of "normal" sexuality, it emphasizes the fact that only one form of sexual identity is right or wrong for a particular individual.

14. Conflict theorists view sexuality and sexual relations as _____.
    a. linked to other forms of subordination that associate sex with power
    b. contributing to the stability of social institutions such as the family
    c. natural behavior that manifests itself similarly across societies
    d. none of these choices

15. Which of the following statements about teen pregnancy in the United States is accurate?
    a. Seventy-eight percent of teen pregnancies in the U.S. are unplanned.
    b. Teen pregnancy rates have increased in the last decade to reach record highs.
    c. The teen pregnancy rate is so high in the U.S., compared with other developed countries, because U.S. teens are more sexually active.
    d. all of these choices

16. Which of the following Supreme Court cases first defined the use of birth control as a right, rather than a crime, in the United States?
    a. Roe v. Wade (1973)
    b. Eisenstadt v. Baird (1972)
    c. Griswold v. Connecticut (1965)
    d. Roberts v. Food and Drug Administration (1960)

17. According to the social construction perspective, which of the following identities is learned through the socialization process?
    a. heterosexuality
    b. homosexuality
    c. bisexuality
    d. all of these choices

18. Based on his research about sexuality on TV talk shows, Joshua Gamson concluded that _____.
    a. television talk shows tend to feature those who least conform to the dominant sexual value system
    b. showcasing sexual nonconformity in a way that may seem freakish, foul-mouthed, and "abnormal" presents a distorted image of gay life, thereby giving legitimacy to those who think that lesbian, gay, bisexual, transgender, and other diverse sexual lifestyles are deviant
    c. although the shows present a distorted image of gay life, they also make diverse sexual identities public, thereby making sexual nonconformity less shocking and thus, in the long run, more acceptable
    d. all of these choices

19. Eugenics refers to the _____.
    a. scientific study of human sexual response and behavior
    b. development of reproductive technology that allows previously infertile couples to have children
    c. application of scientific principles of genetic selection to efforts aimed at perfecting the human species
    d. development of medical technology that improves people's sensitivity to sexual stimulation and heightens their sexual pleasure

20. Research on exposure to pornography in laboratory settings shows that _____.
    a. after exposure to violent pornography, men are more likely to see victims of rape as responsible for their assault, less likely to regard them as injured, and more likely to accept the idea that women enjoy rape
    b. exposure to any kind of pornography increases sexual promiscuity and sexual deviance
    c. exposure to violent pornography almost always leads men to rape or sexually assault women
    d. none of these choices

21. Which of the following statements about abortion in the United States is (are) **true**?
    a. Prior to 1973, only married women had a legal right to abortion in the United States, but after 1973, unmarried women were granted the same right.
    b. The abortion rate has decreased steadily since 1973 when the United States Supreme Court issued a new decision placing more restrictions on the right to abortion.
    c. The United States government has established different criteria for determining if abortion is allowed in the first, second, and third trimesters of pregnancy.
    d. all of these choices

22. The birth control pill _____.
    a. is one of the most widely used forms of contraception in the United States
    b. was approved by the federal Food and Drug Administration (FDA) in 1950
    c. has been legally available to women regardless of their marital status since 1960
    d. all of these choices

True-False Questions

1. Sociologists agree that men have a stronger sex drive than women.
    TRUE or FALSE

2. Tolerance for gay and lesbian relationships varies significantly in different societies around the world.
    TRUE or FALSE

3. Research on sexual identity indicates that once people have engaged in lesbian or gay behavior they inevitably adopt a formal definition of themselves as lesbian or gay.
    TRUE or FALSE

4. Barbara Risman and Pepper Schwartz's review of studies exploring teen sexuality in the 1990s concluded that the sexual behavior of boys was becoming more like girls.
    TRUE or FALSE

5. Most sociological research concerning sexuality in the United States relies on survey data.
    TRUE or FALSE

6. A major factor contributing to the spread of AIDS worldwide is the international sex trade, which includes the sale of sex to promote tourism.
    TRUE or FALSE

7. Homosexual men convicted of child molestation are almost seven times more likely to be imprisoned than are heterosexual men convicted as child molesters with the same criminal record.
    TRUE or FALSE

8. According to national survey research, the proportion of young people who are sexually active has increased, particularly among young women.
    TRUE or FALSE

9. Most states in the U.S. now grant homosexual couples the right to legally register their union and receive the same benefits as heterosexual married couples.
    TRUE or FALSE

10. Of the U.S. public, almost forty percent think abortion should be illegal in all circumstances.
    TRUE or FALSE

Fill-in-the-Blank Questions

1. The _____ refers to the widespread changes in men's and women's roles and a greater public acceptance of sexuality as a normal part of social development.

2. Because lesbians and gays have been denied equal rights and are singled out for negative treatment in society, they are considered a _____ group.

3. According to the _____ perspective, sexual violence such as rape or sexual harassment is the result of power imbalances between women and men.

4. Despite social movement activity to secure equal rights for gay and lesbian people, the fear and hatred of homosexuals, or _____ , is rampant in American culture.

5. Sexual _____ are reflected in personal relationships where one person is powerless or is defined as the property of someone else, and in the sexual exploitation of women in society through pornography and prostitution.

Essay Questions

1. Discuss the ways in which the cultural and social bases of sexuality are revealed, and provide an example for each feature that you identify.

2. Discuss how the Internet has contributed to greater sexual freedom at the same time it has increased the dangers associated with sexual behavior.

3. Explain why the *international sex trade* flourishes in countries such as Thailand.

4. Discuss the debate over the use of the term sexual orientation versus sexual preference.

Solutions

PRACTICE TEST

Multiple Choice Questions

1. D, 296-7 From a sociological perspective, little in human behavior is purely natural. Sexual identity is only one component of the self, and it continues to develop over one's life course, rather than being fixed at birth. Sexual identity develops within a particular cultural, social, and historical context.

2. B, 307 According to Vasquez, men and women who do not conform to strict social definitions of masculinity and femininity are perceived as engaging in gender betrayal, which increases the likelihood that they will be victims of hate crimes.

3. D, 313 The widespread changes in men's and women's roles, a greater public acceptance of sexuality as a normal part of social development, an increase in the use of birth control, and an increase in the incidence of premarital sex for men and women are all consequences of the sexual revolution.

4. D, 302 Sexual politics is linked to the status of women in society, the high rates of violence against women and sexual minorities, and the privilege and power accorded to heterosexuals, among other things.

5. D, 302-3 The patriarchal nature of the priesthood, the sense of secrecy in the private relationship between a parishioner and a priest, and the Catholic Church's hierarchy of power and definition as sacred all contribute to a climate in which unreported abuse can occur.

6. C, 304 Poor women and women of color are most vulnerable to sexual exploitation.

7. D, 304 The cross-national study of sexual jealousy found significant differences in the degree of jealousy.

8. D, 300-1 Sexual identity is socially constructed, learned, and develops through social experience. Coming out is usually not the result of a single homosexual experience, and people do not necessarily move predictably through certain sequences in developing a sexual identity.

9. D, 297 Young children learn appropriate behavior for their gender through sexual scripts.

10. A, 299 Laumann and colleagues conducted a comprehensive survey of the sexual practices of the American public in the early 1990's. The results indicate that young people are becoming sexually active earlier; the proportion of young people who are sexually active has increased, particularly among women; Americans are not very well informed about sex; a significant number of people have extramarital affairs; few people have just one sexual partner in their lifetimes; a considerable number of people are lesbian or gay; and sex is relatively frequent for those who are sexually active.

11. A, 300-1 The percentage of sexually active teens dropped from the early 1990s, and rates of teen pregnancy, abortions, and sexually transmitted diseases decreased.

12. A, 300-1 Barbara Risman and Pepper Schwartz's review of the various studies exploring teen sexuality in the 1990s concluded that the sexual behavior of boys was becoming more like girls. The number of high school boys who were virgins increased, and the sexual activity of African American girls decreased to be closer to that of White and Hispanic girls' behavior which did not increase,.

13. C, 302 Queer theory interprets the various dimensions of sexuality as thoroughly social and constructed through institutional practices. it emphasizes that sexual identity evolves and can change over the life course.

14. A, 300 Conflict theorists see sexuality as part of the system of power relations and economic inequality in society, thus, sexual relations are linked to other forms of subordination. For example, the international sex trade is related to women's high rates of poverty and low social status in many nations around the world.

15. A, 310-12 Teen pregnancy rates in the U.S. have decreased in the last decade, but are still higher than in other developed countries despite the fact that the level of sexual activity is about the same. About 78% of teen pregnancies in the U.S. are unplanned.

16. C, 308 The FDA first approved the birth control pill in 1960. The Supreme Court case, *Griswold v. Connecticut*, first defined contraception as a right, not a crime, in 1965, but restricted this right to married people. In 1972, the court extended this right to unmarried people in *Eisenstadt v. Baird*. In 1973, the court ruled that women have a legal right to abortion in *Roe v. Wade*.

17. D, 300 The social construction perspective interprets sexual identity as learned, not inborn. All sexual identities develop through social experiences within particular social contexts.

18. D, 304-5 Gamson concluded that the result of talk shows presenting a distorted view of gay life presented a dilemma: while in the short run, the distortion gives legitimacy to those who see gay life as deviant, in the long run, those portrayals open up public space, where challenges to sexual conformity can transform the ordinarily fixed boundaries between gay and straight, normal and "queer."

19. C, 309 Eugenics sought to apply scientific principles of genetic selection to "improve" the human race. This philosophy was criticized as being explicitly racist.

20. A, 310 Exposure to violent pornography does produce changes in sexual attitudes, but there is no evidence that it changes sexual behavior, or that exposure to any kind of

pornography increases sexual promiscuity or sexual deviance.

21. C, 309-10 The right to abortion was first established in constitutional law by the Supreme Court case, *Roe v. Wade* (1973). The court ruled that at different trimesters during a pregnancy, separate but legitimate rights collide, including the right to privacy, the right of the state to protect maternal health, and the right of the state to protect developing life. Thus, the government established different criteria for legal access to abortion depending on the stage of the pregnancy.

22. A, 308-9 The FDA first approved the birth control pill in 1960, and today, it is one of the most widely used forms of contraception. The pill was not legally available to unmarried women until 1972.

True-False Questions

1. F, 303 Men do not have a stronger sex drive than women.

2. T, 304 Tolerance for gay and lesbian relationships varies significantly in different societies around the world.

3. F, 300 Some people may engage in lesbian or gay behavior but not adopt a formal definition of themselves as lesbian or gay.

4. T, 300-1 Risman and Schwartz's review of studies exploring teen sexuality in the 1990s concluded

that the sexual behavior of boys was becoming more like girls.

5. T, 299 What we know about sexual behavior is typically drawn from surveys, such as the one conducted by Laumann and colleagues in the early 1990s. Most surveys ask about attitudes, not actual behavior.

6. T, 305 The international sex trade is used to promote tourism and cater to business and military men in countries such as Thailand. It has been strongly

implicated in the spread of AIDS worldwide.

7. T, 306 Homosexual men convicted of child molestation are almost seven times more likely to be imprisoned than are heterosexual men convicted of child molestation with the same criminal record.

8. T, 299 Results of national surveys indicate that the proportion of young people who are sexually active has increased, especially among young women.

9. F, 307-8 Heterosexual couples enjoy institutional privileges such as the right to marry and have mutual employee health benefits, options not usually available to gay and lesbian couples in the United States.

10. F, 309-10 Of the U.S. public, only 20% think abortion should be illegal in all circumstances, 37% think it should be legal in all circumstances, and 39% think it should be legal in a few circumstances.

Fill-in-the-Blank Questions

1. sexual revolution, 313
2. minority, 307-8
3. conflict, 300
4. homophobia, 305
5. politics, 302

Essay Questions

1. Social Construction of Sexuality. Six ways in which sexuality is socially define and patterned are discussed on p. 296-298 in the text. Student responses and examples should reflect an understanding of at least some of these ways in which sexuality is socially constructed.

2. Sex and the Internet. The impact of the Internet on sexuality is discussed on p. 313-314 in the text. Student responses should demonstrate both how the Internet has contributed to greater sexual freedom and how it has increased the dangers associated with sexual behavior.

3. International Sex Trade. The international sex trade is discussed on p. 304-305 in the text. Student responses should reflect an understanding of the roles played by developed and underdeveloped countries in this global trade.

4. Sexual Orientation. Sexual orientation is discussed in the text on p. 300. Student responses should demonstrate an understanding of the term sexual orientation, and what meanings it conveys that are not conveyed by the term sexual preference.

CHAPTER 12

FAMILIES AND RELIGION

BRIEF CHAPTER OUTLINE

Defining the Family

Comparing Kinship Systems

Extended and Nuclear Families

Sociological Theory and Families

Functionalist Theory and Families

Conflict Theory and Families

Feminist Theory and Families

Symbolic Interaction Theory and Families

Diversity Among Contemporary American Families

Female-Headed Households

Married-Couple Families

Stepfamilies

Gay and Lesbian Households

Singles

Marriage and Divorce

Marriage

Divorce

Family Violence

Changing Families in a Changing Society

Global Changes in Family Life

Families and Social Policy

Defining Religion

The Significance of Religion in the United States

CHAPTER FOCUS

This chapter focuses on families and religion as social institutions, exploring the major issues confronting both institutions globally and in the U.S.

QUESTIONS TO GUIDE YOUR READING

1. How do sociologists define the family, what are the main features of the family as an institution, and what are the different kinds of kinship systems and families?

2. What are the major sociological theories of the families, and what are their relative merits.

3. What are the challenges confronting the major family forms in the U.S.?

4. What are the major features of U.S. marriage and divorce patterns, and what factors contribute to divorce?

5. What are the patterns and causes of all forms of family violence?

6. What are the major changes and challenges confronting family life worldwide, and what social policies could effectively respond to those challenges?

7. How do sociologists define and study religion?

8. What is the significance of religion in American society, including the dominance of Christianity and the degree of religiosity?

9. What are the diverse forms of religion globally and in the U.S., what are the roots of religious extremism, and what is the role of religion in social change?

10. How do the major sociological theories approach the study of religion?

11. What are the major types of religious organization, and what are the characteristics of the processes of religious socialization and conversion?

KEY TERMS

(defined at page number shown and in glossary)

antimiscegenation laws 319	**bilateral kinship 319**
charisma 346	**church 345**
collective consciousness 341	**cult 346**
endogamy 319	**exogamy 319**
extended families 319	**family 318**
Family and Medical Leave Act (FMLA) 335	**kinship system 318**
matriarchal religion 340	**matrilineal kinship 319**
monogamy 319	**monotheism 340**
nuclear family 320	**patriarchal religion 340**
patrilineal kinship 319	**polygamy 319**
polytheism 340	**profane 337**
Protestant Ethic 342	**religion 337**
religiosity 339	**religious extremism 344**
ritual 341	**sacred 337**
sect 346	**secular 338**
totem 338	**transnational family 335**

CHAPTER OUTLINE

I. The *family ideal*, represented by a father employed as the major breadwinner and a mother at home raising children, has long been defined within dominant American culture as the family to which we should all aspire, yet few families conform to this ideal today. It is within the family that people are first socialized into religion. Religion has a profound impact on both society and human behavior

II. **DEFINING THE FAMILY**

Despite the diversity of family forms, sociologists define the family as a primary group of people—usually related by ancestry, marriage, or adoption—who form a cooperative economic unit to care for any offspring and each other, and who are committed to maintaining the group over time.

A. Comparing Kinship Systems

   Families are part of **kinship systems**—patterns of relationships that define people's family relationships to one another. Kinship systems vary enormously across cultures and over time. Kinship systems are generally categorized by five features:
   • how many marital partners are permitted at one time
   • who is permitted to marry whom
   • how descent is determined and how property is passed on
   • where the family resides
   • how power is distributed

   1. **Polygamy** is the practice whereby men or women have multiple spouses. *Polyandry*, the practice of a woman having more than one husband, is extremely rare. The more common form of polygamy is *polygyny*, where a man has more than one wife, a practice found among a small group of Mormons in the U.S.
   2. **Monogamy**, the practice of forming a sexually exclusive marriage with one spouse at a time, is the most common form of marriage in the United States.
   3. **Exogamy** is the practice of selecting mates from outside one's group. The group may be based on religion, territory, or racial-ethnic identity.
   4. **Endogamy** is the practice of selecting mates from within one's group.
      a. In the U.S., **antimiscegenation laws**, which prohibited marriage between people of different races, were not declared unconstitutional until 1967.
      b. Even if certain forms of marriage are not explicitly outlawed, societies establish norms about who is an appropriate marriage partner. Although interracial marriage has recently increased, it is still infrequent.
   5. Kinship systems shape the distribution of property in society by proscribing how the lines of descent are determined. In **patrilineal kinship** systems, family lineage or ancestry is traced through the family of the father, whereas ancestry is traced through the mother in **matrilineal kinship** systems. In the U.S., there is a **bilateral kinship system**, where descent is traced through both parents.

B. Extended and Nuclear Families
   1. **Extended families** include the whole network of parents, children, and other relatives who form a family unit. For example, among African Americans, *othermothers* are kin who assist mothers with childrearing responsibilities; and among Chicanos, the system of *compadrazgo* includes help from godparents.
   2. The **nuclear family** is one where a married couple resides together with their children. The origin of this family form in Western society is linked to industrialization. With industrialization, paid labor was performed mostly in factories and public marketplaces, resulting in the separation of the family and the workplace and the development of the *family wage system*.
   3. For groups such as Chinese, Mexican, and Korean immigrants, as well as African Americans, the disruptions posed by slavery, migration, and urban poverty have affected how families are formed, their ability to stay together, and the resources they are able to secure to confront problems.

III. SOCIOLOGICAL THEORY AND FAMILIES

The complexity of family patterns makes it impossible to understand families from any single perspective. Sociologists have used four perspectives in their analysis of families [Table 12.1].

A. Functionalist Theory and Families
   1. Functionalist theorists view the family as fulfilling particular societal needs, including socializing the young, regulating sexual activity and procreation, providing physical care for family members, and giving psychological support and emotional security to individuals. Over time, other institutions have begun to fulfill some of the functions originally performed by the family.
   2. Functionalists conceptualize marriage as a mutually beneficial exchange wherein women receive protection, economic support, and status in return for emotional support, sexual intimacy, household maintenance, and production of offspring.
   3. When societies experience disruption and change, social institutions such as the family become disorganized, which weakens the social order.

B. Conflict Theory and Families

   Conflict theorists interpret the family as a system of power relations that both reinforces and reflects the inequalities in society at large. Families socialize children to obey authority and to become good consumers to fit the needs of capitalism.

C. Feminist Theory and Families

   Feminist theory makes gender a central concept in the analysis of the family as a social institution. Influenced by conflict theory, feminist scholars have asserted that the family does not serve the needs of all members equally, because the family is a system of gendered power relations.

D. Symbolic Interaction Theory and Families

   Symbolic interaction theory focuses how people define and understand their family experiences as well as how people negotiate family relationships. This perspective emphasizes the construction of meaning within families and notes that family roles continually evolve as participants actively create family life and relationships.

IV. DIVERSITY AMONG CONTEMPORARY AMERICAN FAMILIES

Families are systems of social relationships that emerge in response to social conditions and that, in turn, shape the future direction of society. Compared to families in the past, families are smaller in size and devote fewer years to childcare, but more years to elder care. Single-parent households, post childbearing couples, gay and lesbian couples, and those without children are increasingly common.

A. Female-Headed Households
   1. One of the most significant changes in family life has been an increase in single parent families, with one-fourth of all children now living with one parent.

2. Women head 88% of single-parent families. The two main reasons for an increase in the number of women heading their own households are the high rates of divorce and teen mothers who do not marry.

3. Sociologists suggest that the economic pressures faced by women in female-headed households puts them under great strain due to poverty, a phenomenon known as the *feminization of poverty*.

4. The number of families headed by single fathers is also increasing, although only 16% of children living with a single parent live with their fathers. Male-headed households are less likely to experience severe economic problems, and single fathers generally get housework and childcare help from other women—girlfriends, daughters, or mothers.

B. Married-Couple Families

Among married-couple families, one of the greatest changes in recent years has been the increased participation of women in the paid labor force. This has created other changes, such as an increase in the number of *commuter marriages*, an arrangement that typically arises when work requires one partner in a dual-earner couple to reside in a different city. Women often experience greater independence and more egalitarian relations within the family during and after such arrangements, despite the financial strains and stress on family relationships.

C. Stepfamilies

Stepfamilies have become more common in the United States, with about 40 percent of marriages involving stepchildren. Both parents and children must learn new roles when they become part of a blended family. Problems of jealousy and competition for time and attention can increase family tensions, which are exacerbated by a lack of clear norms.

D. Gay and Lesbian Households
1. Research indicates that gay and lesbian couples tend to be more flexible and less gender-stereotyped in household roles than heterosexual couples.

2. The balance of power between lesbian couples seems unaffected by the amount of money each partner makes, but the highest earner usually has more power among gay male couples. Patterns of employment do tend to influence parenting roles among lesbian couples, however.

3. Gay and lesbian couples are negatively affected by the denial of benefits and privileges accorded legally recognized marriages. This has fueled the contemporary debate about whether gay marriage should be recognized in law, while there is increasing acceptance of gay and lesbian relationships.

E. Singles

Single people, including those never married, widowed, and divorced, constitute 44 percent of the population today, an increase from 29 percent in 1970. Patterns of establishing intimate relationships have changed in the United States, with *cohabitation* becoming quite common. Additionally, a growing number of people are remaining in their parents' homes for longer periods of time, particularly for economic reasons.

V. **MARRIAGE AND DIVORCE**

The United States has the highest rate of marriage of any Western industrialized nation, but it also has a high divorce rate.

A. Marriage
    1. The U.S. has the highest rate of marriage of any Western industrialized nation.
    2. Gender, income, and occupational level all influence power within marriage, including the division of labor.
    3. Among working couples, 28% share housework, with little difference across social class. Women continue to do far more work at home and have less leisure time than men (the *second shift*). Men are working longer hours, but primarily in paid employment. The majority of women in all social classes experience stress over the amount of work they have to do and their lack of free time.
    4. Children can enrich a marriage, but also create tension; marital satisfaction is higher before and after the parental role.

B. Divorce
    1. The U.S. also has the highest divorce rate. The divorce rate is higher for young couples, in second marriages, among low-income couples, and at the extremes of educational attainment. The divorce rate for African Americans is higher than for Whites.
    2. Women are less likely than men to remarry after divorce.
    3. The continuing rise in life expectancy, the high valuing of individualism, and the change in women's roles have all contributed to the high divorce rate in the U.S. Despite the emotional pain and economic struggle women experience following divorce, most are glad that their marriages ended.
    4. A number of factors influence children's adjustment to divorce. Violence and prolonged discord between the parents before, during, and after divorce increases the likelihood of difficulties for children after the breakup. The absence of fathers and limited child support also contribute to post-divorce problems for children.

C. Family Violence

    Partner Violence. The majority of domestic violence cases go unreported, but it is currently estimated that 25% of women will be raped, assaulted or stalked by an intimate partner. In some instances, wives assault husbands, and partner abuse among gay and lesbian couples appears comparable to abuse among heterosexual couples. Most battered spouses do leave abusive relationships and seek ways to prevent further victimization, but complex factors—including belief that the batterer will change, financial constraints, and mandatory arrest laws—may discourage reporting and keep the battered victim in the relationship. Sociological analyses conclude that women's relative powerlessness in the family is at the root of high rates of violence against women.

    Child Abuse. Child abuse is associated with chronic alcohol use by a parent, unemployment, isolation of the family, and the absence of social supports. Women are as likely as men to be perpetrators.

    Incest. Incest is a certain form of child abuse involving sexual relations between closely related persons, with fathers and uncles being the most likely perpetrators. The most likely families to be affected are those in which the mother is debilitated.

    Elder Abuse. Most of the care of older people in the U.S. is provided informally by families, mostly by women, who often experience stress from the demands of caregiving. Sons are most likely to engage in physical abuse.

VI. CHANGING FAMILIES IN A CHANGING SOCIETY

A. Global Changes in Family Life

The increasing global basis of the economy has facilitated new patterns of work and migration that have created a new family form, the **transnational family**. This is a family form in which one or both parents live and work in one country, while their children remain in their countries of origin.

B. Families and Social Policy

Balancing Work and Family. The Family and Medical Leave Act (FMLA) of 1993 began to address the needs of families confronted with the care of children and other dependents New Social policies are needed to address the needs of diverse families.

Child Care. Almost half of families with children under age 13 have child care expenses, typically taking 9% of their earnings, with single-parent and low-income families spending 16%. *Care work* is increasingly provided to middle- and upper-class families by women of color and immigrant women.

Elder Care. Women, who shoulder the work of elder care, can expect to spend more years caring for an elderly parent than raising their own children.

VII. DEFINING RELIGION

Sociologists study religion as a belief system and a social institution. The belief systems of religion powerfully influence how people see the world. As a social institution, religion is among the most important influences on people's lives. Religious beliefs and practices are related to social factors such as race, class, age, and gender. Sociologists define **religion** as an institutionalized system of symbols, beliefs, values, and practices through which a group of people responds to what they feel is sacred and find answers to questions of ultimate meaning.
- **Religion is institutionalized.**
- **Religion is a feature of groups.**
- **Religions are based on beliefs that are considered sacred, as opposed to profane**. A **totem** is an object or living thing that a religious group regards with special reverence.
- **Religion establishes values and moral proscriptions for behavior.**
- **Religion establishes norms for behavior.**
- **Religion provides answers to questions of ultimate meaning.**

VIII. THE SIGNIFICANCE OF RELIGION IN AMERICAN SOCIETY

A. The Dominance of Christianity

Despite the constitutional principle of the separation of church and state, Christian beliefs and practices dominate American culture. An example of this is that Christian traditions are publicly observed through the designation of national holidays in the United States.

B. Measuring Religious Faith

Religiosity refers to the intensity and consistency of a group's or person's faith, which is measured by asking people about their religious beliefs and tallying membership in religious organizations and attendance at religious services. Over 50 percent of the population of the United States identify themselves as Protestant.

C. Forms of Religion

Religions are categorized according to the specific characteristics of the faiths and how religious groups are organized.

1. **Monotheism** is the worship of a single god, as in Christianity, Judaism, and Islam, while **polytheism** is the worship of more than one deity, as in Hinduism.
2. **Patriarchal religions** are those in which the beliefs and practices are based on male power and authority. **Matriarchal religions** are based on the centrality of female goddesses, who are often seen as the source of food, love, and nurturance, and may serve as emblems of the power of women.

IX. SOCIOLOGICAL THEORIES OF RELIGION

A. Emile Durkheim: The Functions of Religion
    1. Durkheim argued that religion is functional for society because it reaffirms the social bonds that people have with each other, which creates social cohesion.
    2. Religious **rituals** are symbolic activities that express a group's spiritual conviction, such as a pilgrimage to Mecca.
    3. Durkheim believed that religion binds individuals to society by establishing what he called a **collective consciousness**, or the body of beliefs that are common to a community and that give people a sense of belonging.

B. Max Weber: The Protestant Ethic and the Spirit of Capitalism

    Max Weber posited that the Protestant faith supported the development of capitalism in Western Europe and the United States by professing a belief in predestination, the idea that one's salvation is predetermined by god. Because material success was viewed as a gift from god, rather than something a person could earn, successful people appeared to be favored by god. Weber noted that the core features of the **Protestant ethic**, hard work and self-denial, led not only to religious salvation, but also to the accumulation of capital.

C. Karl Marx: Religion, Social Conflict, and Oppression
    1. Karl Marx viewed religion as a form of *false consciousness* and a tool for class oppression. He argued that oppressed people develop religion to soothe their distress, which prevents them from rising up to challenge the oppressive system.
    2. Marx defined religion as an *ideology*, or belief system that legitimates the existing social order. This is reflected in the historical situation of Christian principles being invoked by slave owners to justify slavery in the United States.
    3. It is important to note that religion can also be the basis for liberating social change, as demonstrated by the U.S. Civil Rights Movement and Latin American liberation movements.

D. Symbolic Interaction: Becoming Religious

Becoming religious is the result of religious socialization that can be informal or formal. Initial religious socialization usually occurs within the family, although for others, it involves a gradual or more dramatic process of conversion. Sociologists counter the popular notion of "brainwashing" by emphasizing that conversion is linked to shifting patterns of association and that people are active participants in the conversion process. Religious conversion is a social process, usually entailing three phases: disruption in previous life experience leading to withdrawal and some loss of autonomy, creation of an emotional bond with group members and a weakening of former bonds, and a period of intense interaction with the new group.

X. DIVERSITY AND RELIGIOUS BELIEF

Around the world, Christianity and Islam have the largest memberships [Figure 12.8]. In the United States, religious identification varies by race-ethnicity, gender, age, income and education level, and political affiliation. For example, young people are more likely than older people to express no religious preference, and considerably more women than men report that religion is very important to them [Figure 12.6].

A. The Influence of Race and Ethnicity.

For some racial-ethnic minority groups, religion can be a defense against racism (as among the Black Muslims), and churches have historically been among the most important institutions within African American and Latino communities. Among Asian Americans, which enjoy diverse religious traditions, as well as among other immigrant communities, the discontinuity with a religious past brought on by cultural assimilation can be a source of intergenerational tension.

B. Religious Extremism.

Religious extremism refers to actions and beliefs that are driven by high levels of religious intolerance, and that see the world in simplistic categories that fuel hate and conflict. While popularly associated with the Middle East, all religions can develop extremism. Religious extremism is learned, typically emerging from countries where people are very poor and without access to education, and where there is a lack of modernism. Religious extremist movements tend to be highly patriarchal.

XI. RELIGIOUS ORGANIZATIONS

Sociologists define religious organizations using a classification system of three *ideal types*.

A. **Churches** are formal organizations that tend to define themselves, and be seen by society, as the primary and legitimate religious organizations. They are often organized as complex bureaucracies and membership is renewed as children of existing members are brought up in the church.
   1. **Sects** are groups that have broken off from an established church, such as the Shakers' departure from the Society of Friends (also known as Quakers). Sects tend to focus less on the organization per se and more on the purity of members' faith, admitting only truly committed members to the group.
   2. **Cults**, similar to sects in their intensity, are religious groups devoted to a specific cause or charismatic leader. Cult leaders tend to have great **charisma**, a quality attributed to individuals who are believed to have special powers. Cults are close-knit communities that attract people longing for personal attachments.

XII. RELIGION AND SOCIAL CHANGE

Religion remains an important social institution that, like other aspects of society, is becoming more commercialized. The role of religious organizations in social change has recently become a question of public policy and debate, especially related to funding of faith-based organizations for humanitarian work. Religion continues to have an important role in liberation movements around the world. The role of women is also changing in most religious organizations.

INTERNET EXERCISES

1. The *Family Violence Prevention Fund* homepage (endabuse.org) provides easy access to its multiple programs and informational links. The programs of the FVPF include public education, healthcare, workplace, children, justice, international, immigrant women, public policy and economic independence. Choose one program area, and explore the analysis, policies and best practices, and write a report evaluating the approaches advocated from a sociological perspective.

2. The *Harvard Pluralism Project* (www.fas.harvard.edu/pluralism/) seeks "to help Americans engage with the realities of religious diversity through research, outreach, and the active dissemination of resources." There are several features on the site including Resources by Religious Tradition, Resources by State (with an interactive map), and Religious Diversity News Headlines. Students can also browse through Religion and Politics 2004: Multifaith Resources and report on what they learned about the way religions sought to engage in and influence the 2004 elections.

INFOTRAC EXERCISES

Conduct a keyword search using InfoTrac College Edition to extend the discussion in DOING SOCIOLOGICAL RESEARCH: *Men's Caregiving* (p. 332-3).

1. Keyword: Family division of labor. This search leads to articles contrasting cohabitation and marital unions, gender expectations and entitlement.

2. Keyword: Gender and housework. This search yields articles on men doing housework in the absence of women, gender ideologies, relative resources, social class, race and ethnicity, cross-national studies, dual-earner couples, and contrasts between Black and White couples.

3. Keyword: Men and child care. This search leads to studies in the U.S., Russia, and Europe.

4. Keyword: Men and nurturing. Searching with this keyword leads to articles on Asian-American men, older men, single fathers, and media depictions of nurturing men.

PRACTICE TEST

Multiple Choice Questions

1. From a sociological perspective, which of the following statements about religion is **false**?
   a. Religious faith provides a source of cohesion for members of society, but it also serves as a source of conflict in society.
   b. Religion cannot be studied as a social institution because it is based on faith rather than objective principles.
   c. A manifest function of religion is to provide a sense of ultimate meaning and purpose.
   d. Religion establishes values and norms that restrict people's behavior in society.

2. Religiosity refers to _____.
   a. the socialization mechanisms that teach people moral values and guidelines for ethical behavior
   b. the degree of influence religious organizations have on a society's political system
   c. whether one believes in a single, all powerful deity or multiple deities
   d. the intensity and consistency of practice of a person's faith

3. Which of the following religions is polytheistic?
   a. Christianity
   b. Islam
   c. Judaism
   d. none of these choices

4. Religions based on the centrality of female goddesses are classified as which type of religion?
   a. Monotheistic
   b. Matriarchal
   c. Afrocentric
   d. Patriarchal

5. The profane is anything that _____.
   a. is supernatural
   b. excites awe and reverence
   c. is regarded as part of the ordinary world
   d. can be handled only by people who have been officially "blessed" or "ordained"

6. Symbolic activities that express a group's spiritual conviction, such as a pilgrimage to Mecca, are religious _____.
   a. orientations
   b. ideologies
   c. doctrines
   d. rituals

7. According to Emile Durkheim, religion is _____.
   a. a system of socially constructed beliefs subject to individual interpretation
   b. no longer important in industrialized societies
   c. functional for society because it reaffirms the social bonds that people have with each other, thereby fostering cohesion
   d. an ideology that legitimates an oppressive social order and supports the status quo

8. Which theorist argued that the Protestant ethic encouraged and supported the development of capitalism in the United States and Western Europe?
   a. Karl Marx
   b. Max Weber
   c. Sigmund Freud
   d. Emile Durkheim

9. With which type of kinship system in the United States are "othermothers" and *compadrazgo* associated?
   a. nuclear
   b. bilateral
   c. extended
   d. patrilineal

10. When measured in terms of followers, the largest religion in the world is _____.
    a. Islam
    b. Hinduism
    c. Buddhism
    d. Christianity

11. The most common marriage pattern in the United States and other industrialized nations is _____.
    a. exogamy
    b. polygamy
    c. monogamy
    d. patrilineality

12. Research on decision making and household labor in married couple families reveals that _____.
    a. regardless of the amount of money that wives contribute to the household, they have more influence over household decisions than do their husbands
    b. the men and women who are most likely to develop an egalitarian division of household labor when they get married are those whose own parents divided housework equitably
    c. because women's paid labor force participation has increased so much in recent years, married men now do about half of all housework
    d. all of these choices

13. Which of the following statements about female-headed households is (are) **true**?
    a. The high rate of pregnancy among unmarried teens and the high divorce rate have led to the growing number of women heading their own households.
    b. Divorced women experience a substantial decline in income, and most receive little or no financial support from their former husbands.
    c. Social problems associated with female-headed households, like delinquency, school dropout rate, and poor self-image, are caused not by the absence of men in the family, but rather by economic strains.
    d. all of these choices

14. Functionalist theorists view families as _____.
    a. systems of power relationships that reinforce the inequalities in society at large
    b. organized around a harmony of interests and beneficial to all of the members
    c. gendered institutions that reflect the gender hierarchy in society at large
    d. groups in which people interact and negotiate relationships with others

15. Symbolic interaction theorists view families as _____.
    a. systems of power relationships that reinforce the inequalities in society at large
    b. organized around a harmony of interests and beneficial to all of the members
    c. gendered institutions that reflect the gender hierarchy in society at large
    d. groups in which people interact and negotiate relationships with others

16. Which perspective interprets people as moving into religious cults gradually, especially when they are open to changes in the social environment and seek meaningful personal attachments?
    a. symbolic interaction theory
    b. functionalist theory
    c. liberation theology
    d. conflict theory

17. Which of the following statements about single-parent households in the United States is **true**?
    a. Fifty percent of all American children currently live with only one parent.
    b. Approximately one-fourth of all single parent households are headed by men.
    c. A primary cause for the increase in single-parent households is the high rate of divorce.
    d. all of these choices

18. Which of the following about the divorce rate in the United States is accurate?
    a. The divorce rate is higher for Whites than for African Americans.
    b. The divorce rate is highest for first marriages, because people learn from their mistakes.
    c. The divorce rate is highest among upper-class couples.
    d. none of these choices

19. Karl Marx argued that religion is a form of false consciousness because:
    a. subordinate groups internalize religious ideology that justifies an oppressive social order.
    b. dominant religious ideology strongly encourages subordinate groups to seek justice and equality by challenging oppression.
    c. religion binds individuals to each other and to society, thereby enhancing cohesion and social harmony.
    d. religious worship no longer has strong meaning for people in modern societies because it takes place in large, formal organizations rather than small, intimate gatherings.

20. The largest religious group people in the United States identify as their religious affiliation is _____.
    a. Judaism
    b. Islam
    c. Protestantism
    d. Roman Catholicism

21. Cults tend to form around leaders with great _____, a quality attributed to individuals believed by their followers to have special powers.
    a. conservatism
    b. materialism
    c. liberalism
    d. charisma

22. Groups that have broken off from an established church, and which place less emphasis of organization and more emphasis on the purity of members' faith, are called _____.
    a. support groups
    b. temples
    c. totems
    d. sects

23. A(n) _____ is an object, such as a statue of Buddha, that a religious group regards with special awe and reverence.
    a. totem
    b. trinket
    c. ideology
    d. theology

24. Which industrialized nation provides the shortest length of maternity leave and the lowest percentage of wages paid during the covered period?
    a. United States
    b. Canada
    c. Russia
    d. Japan

25. Which of the following statements about elder care in the U.S. is **true**?
    a. Family members provide only about 10 to 20 percent of long-term care for the elderly.
    b. Since Medicare covers long-term care expenses, most families place their elders in nursing homes, or care for them at home with financial support from the state.
    c. As more women are working in the labor force, the family contribution to elder care has shifted, with a great burden assumed by men.
    d. none of these choices

True-False Questions

1. Antimiscegenation laws, which prohibit marriage between people of different racial groups, were not declared unconstitutional until 1867.

   TRUE or FALSE

2. The influence of evangelical religious groups on the political system in the United States has decreased considerably in recent years.

   TRUE or FALSE

3. The United States leads the world not only in the number of people who marry, but also in the number of people who divorce.

   TRUE or FALSE

4. Religious organizations today reflect considerable gender equality, with women accorded full participation and ordination as clergy in most faiths.

   TRUE or FALSE

5. Large numbers of African Americans have recently become Catholics because they are drawn to the church's emphasis on self-reliance, celebration of African identity, and prohibitions against gambling and using drugs or alcohol.

   TRUE or FALSE

6. Judaism has such enormous social significance because it is the second largest religion in the world.

   TRUE or FALSE

7. According to feminist theory, marriage is a mutually beneficial arrangement wherein women receive protection and financial support in exchange for providing emotional intimacy, child care, and housework.

   TRUE or FALSE

8. Research indicates that children who are raised by gay or lesbian parents experience significantly more psychological problems than do children raised by heterosexual parents.

   TRUE or FALSE

9. According to Khanna and colleagues' research on interracial dating and marriage in the United States, the families of White students react with hostility to the news that their relative is involved in an interracial relationship, but the families of Black students who date interracially are generally supportive of their relative's involvement in the relationship.

TRUE or FALSE

10. The children who are most likely to have ongoing emotional difficulties as the result of their parents' divorce are those whose parents maintain joint custody.

TRUE or FALSE

Fill-in-the-Blank Questions

1. Science generates _____ beliefs based on logic and rational observations, rather than faith.

2. Hinduism is a _____ religion because it is based on the belief that there are millions of gods and demons, rather than one powerful god.

3. Christianity is categorized as a _____ religion because its doctrine and practices are based on male power and leadership.

4. In a _____ family, one or both parents live in one country while their children remain in their countries of origin.

5. In a _____ family, a married couple resides together with their children.

Essay Questions

1. Summarize the results of research on the consequences of divorce for children, noting which factors promote or impede their adjustment.

2. Discuss the factors that contribute to the high divorce rate in the United States.

3. Discuss the factors that influence the patterns of care work in the family.

4. Discuss the roots of religious extremism.

5. Discuss the role of religion in U.S. public life and politics.

Solutions

PRACTICE TEST

Multiple Choice Questions

1. B, 337 From a sociological perspective, religion can be scientifically studied as a social institution. Religion establishes values, proscribes norms for moral behavior, and provides members of society with a sense of meaning. Religion can serve as a source of cohesion, conflict, and social change.

2. D, 339 Religiosity refers to the intensity and consistency of practice of an individual's or group's faith.

3. D, 340 Monotheistic religions such as Christianity, Judaism, and Islam accept the idea that there is a single, all-powerful god. In Hinduism, a polytheistic religion, god is not a specific entity at all.

4. B, 340 Religions based on the centrality of female goddesses are classified as matriarchal, while patriarchal religions are based on male power and authority.

5. C, 338-9 The profane includes anything regarded as part of the ordinary world. The sacred refers to things of a supernatural nature that are regarded with awe and reverence. The handling of sacred objects is often restricted to religious authorities, or may be done by ordinary people if they undergo certain religious rituals.

6. D, 341 Symbolic activities that express a group's spiritual conviction are religious rituals. Totems are objects or living things that religious groups regard with awe.

7. C, 341 According to Durkheim, a functionalist theorist, religion is functional and necessary for all societies because it affirms and strengthens social bonds. Marx, who represents conflict theory, argued that religion is an ideology that legitimates the existing social order, supports the status quo, and benefits the ruling class. Symbolic interaction theory views religion as a system of socially constructed beliefs subject to various interpretations [Table 12.3].

8. B, 342 Max Weber noted that the Protestant ethic encouraged hard work, self-denial, and personal accumulation of wealth, which was instrumental to the development of capitalism.

9. C, 319-20 "Other mothers" and *compadrazgo* provide childcare and other forms of assistance to families organized around extended kinship systems. The nuclear family, by contrast, consists of a married couple and their children.

10. D, 343 Figure 12.5 indicates that Christianity is the largest religion in the world.

11. C, 319 Monogamy, or being married to only one person, is the most common form of marriage in the United States and other Western, industrialized nations.

12. B, 330-1 Research indicates that most married couples do not share housework equally, although fathers' participation in child care has increased, especially when there are very young children in the household. Men and women who had parents who shared housework equitably are more likely to have an egalitarian division of labor in their own households when they grow up and get married.

13. D, 325 The high rate of pregnancy among unmarried teens and the high divorce rate have both contributed to the growing number of women heading their own households. Divorced women experience a substantial decline in income, and most receive little or no financial support from their former husbands. Social problems associated with female-headed households, like delinquency, school dropout rate, and poor self-image, are caused not by the absence of men in the family, but rather by economic strains.

14. B, 321-2 Functionalist theorists view families as cohesive groups organized around a harmony of interests and an efficient division of labor. Conflict and feminist theorists view families as systems of power relationships that reinforce social inequalities, and feminist theorists further emphasize that families are gendered institutions. Symbolic interaction theory focuses on families as groups in which members interact and negotiate relationships with each other.

15. D, 323 Symbolic interaction theory focuses on families as groups in which members interact and negotiate relationships with each other.

16. A, 342-3 Symbolic interaction theory views people as voluntarily moving into cults gradually as they seek a sense of meaning and attachment to other people.

17. C, 325 Although the number of single parent households headed by men has increased, the majority of single parent households are headed by women. The primary causes of single parent households in the United States are childbearing by single teens and the high rate of divorce in this country. One-quarter of American children live with only one parent.

18. D, 331-3 The divorce rate is higher for African-Americans, second marriages, and low-income couples.

19. A, 342 Karl Marx, the founder of conflict theory, argued that religion is a form of false consciousness because subordinate groups internalize dominant ideology, which justifies an oppressive social order.

20. C, 339 Forty-eight percent of people in the United States identify their religious affiliation as Protestant. Jewish people are least represented among these religious groups.

21. D, 346 Cults tend to form around leaders with charisma, a quality attributed to individuals believed by their followers to have special powers.

22. D, 346 Sects, such as the Shakers or the Amish, are groups that break off from established churches. Churches are formal organizations that are usually socially defined as the primary, legitimate religious organizations in American society. Cults, which are religious groups devoted to a specific cause or charismatic leader, are similar to sects in their intensity and concern with members' purity of faith.

23. A, 338 A totem is an object or living thing, such as a statue of Buddha, which religious groups regard with awe and reverence. An ideology is a belief system. Theology refers to the study of religious doctrine.

24. A, 337 As indicated by Table 12.2, the United States provides only 12 weeks of unpaid leave for maternity leave benefits (but not to all employees), the lowest among industrialized nations, and many less developed nations as well.

25. D, 336-7 Family members, most often women, should the burden of the long-term care, mostly as unpaid labor. Medicare does not provide long-term care benefits.

True-False Questions

1. F, 319 Antimiscegenation laws were not declared unconstitutional until 1967.

2. F, 340 Conservative evangelical religious groups have had an increasing influence on American politics in recent years.

3. T, 331 The United States has the highest rates of both marriage and divorce among the industrialized nations.

4. F, 347 Although women's full participation in religious organizations has increased, gender inequality persists. Many religious organizations do not allow the ordination of women clergy.

5. F, 343 Many African Americans have joined the Black Muslims, a religious group that celebrates Black people's African heritage and identity, has strict prohibitions against the use of alcohol and drugs, and proscribes dietary regulations.

6. F, 343 Around the world, Judaism has significantly few members compared to other major religious groups, such as Christianity and Islam [Figure 12.5].

7. F, 322 Functionalist theory views marriage as a mutually beneficial exchange for men and women. Feminist theory views the family as a gendered institution that reflects the pattern of gender inequality in the society at large, which tends to socially and economically disadvantage women [Table 12.1].

8. F, 328 There is no evidence that being raised by gay or lesbian parents increases the chance of a child developing psychological problems. Children of gay or lesbian parents may learn more flexible gender roles and learn to respect differences.

9. F, 320-1 Interracial marriage in the United States is still relatively uncommon. Khanna and colleagues' research indicates that both Blacks and Whites in interracial dating relationships report negative responses from their families, although the majority of these are not hostile reactions.

10. F, 332-3 Research indicates that those children who have the most difficulty adjusting to their parents' divorce are children who live in households where marital conflict was high before, during, and after the divorce.

Fill-in-the-Blank Questions

1. secular, 338

2. polytheistic, 340

3. patriarchal, 340

4. transnational, 335

5. nuclear, 320

Essay Questions

1. Effects of Divorce on Children. The consequences of divorce for children is discussed on p. 332-333 in the text. Student responses should demonstrate awareness of the research on the effects of divorce on children of different ages, including the factors that promote or impede their adjustment.

2. Divorce in the U.S. The factors that contribute to the high divorce rate in the United States are discussed on p. 331-333 in the text. Student responses should demonstrate familiarity with the divorce rates and how they are influenced by various social factors.

3. Carework. Carework in the family is discussed on p. 335-337 in the text. Student responses should demonstrate familiarity with the various forms of carework and the factors that influence the patterns of this work in the family.

4. Religious Extremism. The roots of religious extremism are discussion on p. 344-345 in the text. Student responses should demonstrate familiarity with social factors that influence the rise of religious extremism.

5. Religion in U.S. Politics. The role of religion in U.S. public life and politics is discussed on p. 346-347 in the text. Student responses should demonstrate familiarity with the constitutional issues and the recent debate over faith-based initiatives receiving governmental support.

EDUCATION AND HEALTH CARE

BRIEF CHAPTER OUTLINE

CHAPTER FOCUS

This chapter examines the social institutions of education and healthcare. Education is considered in terms of its functions and conflicts, its relationship to occupation, income, and social mobility, its impact on social inequalities, and controversies over testing, tracking, teacher expectancy, stereotype threat, and gender. The emergence of modern medicine and the patterns of health, diversity and social inequality in the U.S. are considered, along with the three major theoretical perspectives. The health care crisis in the U.S., including costs and medical malpractice, are explored, together with the response of HMOs and a consideration of the universal health care debate.

QUESTIONS TO GUIDE YOUR READING

1. What are the functions of education in the U.S., and what are the societal conflicts that are reflected in the American school system?

2. What is the relationship between education and occupation, income, and social mobility?

3. How does education both reduce and perpetuate or increase inequalities?

4. What are the controversies over ability and achievement testing?

5. What are the impacts of tracking, teacher expectancy, stereotype threat, and gender on student achievement?

6. What are the major features of the emergence of modern medicine?

7. What are the relationships among health, diversity and social inequality?

8. What are the patterns of AIDS and STDs and the role of stigma in their spread and treatment?

9. How do the three major theoretical perspectives analyze health care?

10. What are the major dimensions and causes of the health care crisis in America and the proposals to address it?

KEY TERMS

(defined at page number shown and in glossary)

achievement test 358	anorexia nervosa 366
cognitive ability 358	cognitive elite 360
credentialism 354	defensive medicine 376
epidemiology 367	health maintenance organization (HMO) 376
latent function 354	managed care 376
Medicaid 375	Medicare 375
predictive validity 359	self-fulfilling prophecy 362
social epidemiology 367	standardized ability test 358
stereotype threat effect 364	stigma 370
teacher expectancy effect 362	tracking 361

CHAPTER OUTLINE

I. **SCHOOLING AND SOCIETY**

  A. <u>The Rise of Education in the United States</u>

   Education in society is concerned with the systematic transmission of society's knowledge. *Schooling* refers to the formal, institutionalized aspects of education. By 1900, compulsory education was established by law, but state laws often excluded minorities like Black Americans, Hispanics, American Indians, and Chinese immigrants. Today, almost 90 percent of those under 35 in the U.S. have at least a high school diploma.

  B. <u>The Functionalist View of Education</u>

   All known societies have some sort of educational institution, which is generally large and highly formalized in industrialized societies such as the United States. Functionalist theory argues that education accomplishes certain functions for society.

   1. *Socialization* occurs as the cultural heritage is transmitted from one generation to another. In addition to teaching a variety of skills and knowledge, schools also inculcate values such as loyalty and punctuality.

2. *Occupational training* is provided by schools. Modern industrialized societies especially need a system to train people for jobs. Most jobs today require at least a high school education, and many professions require a graduate degree.

3. *Social control* is also provided by the educational institution. This indirect, non-obvious consequence of schools is called a **latent function**. One perceived benefit of compulsory education in the late 19th century was to keep kids off the streets and out of trouble as crime, overcrowding, and other problems intensified with urbanization and immigration.

C. The Conflict View of Education

Conflict theory emphasizes the disintegrative and disruptive aspects of education by focusing on competition between groups for power, income, and social status. The unequal distribution of education can allow educational level to be used as a tool for discrimination through the process of **credentialism**, or the insistence upon educational credentials for their own sake, even if the credentials bear little relationship to the intended job.

D. The Symbolic Interactionist View of Education

Symbolic interaction theory focuses on what emerges from the process of interaction between school staff and students during the schooling experience. For example, the expectations a teacher has for a student can create the very behavior that is expected, such as high or low test performance, through the *expectancy effect*.

II. **DOES SCHOOLING MATTER?**

A. Effects of Education on Occupation and Income

1. One way that sociologists measure a person's social class or socioeconomic status (SES) is to determine the person's amount of schooling, income, and type of occupation, which are *indicators* of SES.

2. In the general population, there is a strong correlation between level of formal education and *occupational prestige*, or the social value attributed to jobs.

3. Gender also influences the relationship between education and income, with women's average income being less than men's at each education level.

4. At least one researcher argues that a serious mismatch exists between the skills students learn and the skills required to enter the job market.

B. Effects of Social Class Background on Education and Social Mobility

Education has traditionally been viewed as the principal route to upward social mobility in the United States. However, research demonstrates that the effect of education upon a person's eventual job and income depends to a great extent upon the social class into which the person was born.

C. Education, Social Class, and Mobility Seen Globally

Some people have argued that there is more occupational and income mobility in the U.S. than in other countries because of the American educational system, but the degree of mobility is relatively limited. For example, there is a dramatic increase in average SAT (Scholastic Aptitude Test) scores as family income increases [Table 13.3]. This increases the likelihood that students from higher-income families will get into the best colleges.

III. **EDUCATION AND INEQUALITY**

Education reduced many but not all social inequalities since the introduction of compulsory education at the turn of the 20th century. For example, the percentage of high school graduates has risen among Whites and minorities, both male and female, and there has been an overall increase in minorities' and women's college attendance.

A. Cognitive Ability and Its Measurement
　　1. The American educational system relies heavily on **standardized ability tests** (like the SAT or ACT) to measure "intelligence," as opposed to **achievement tests**, which measure what has been learned.
　　2. There are three major criticisms of the use of standardized tests as measures of **cognitive ability**.
　　　　a. They tend to measure only limited ranges of ability, such as quantitative or verbal aptitude, while ignoring other cognitive endowments, such as creativity.
　　　　b. These tests possess a degree of *cultural* and *gender bias* that may perpetuate social, economic, and educational inequality.
　　　　c. The **predictive validity** of these tests, or the extent to which the tests accurately predict later college grades, is compromised for minorities and women. In fact, SAT scores are only moderately accurate predictors of college grades, even for White students.

B. Ability and Diversity
　　1. On average, Whites score higher on standardized ability and achievement tests than minorities; men score higher than women, especially on the math portion; and the higher a person's social class, the higher the test score.
　　2. There is no evidence that *between-group* differences are genetically inherited. Certain *within-group* differences may reflect genetic differences among individuals within the same racial or ethnic group, but even in this category, the effects of social environment are greater than the effects of genes.
　　3. The gap separating White males from minorities and White women, however, is shrinking, undermining the belief that perpetuates educational and occupational inequality.

C. The "Cognitive Elite" and *The Bell Curve* Debate
　　1. The *Bell Curve*, published in 1994, created an intense debate about the nature of intelligence. Authors Herrnstein and Murray argued that not only does the distribution of intelligence in the general population closely approximate a bell-shaped curve, or *normal distribution*, but that there is one basic, fundamental kind of intelligence.
　　2. Drawing on studies of identical twins, the authors concluded that intelligence has 70 percent *heritability*, while only 30 percent of the difference in intelligence throughout the population is determined by environment.
　　3. The authors argued that the lower classes are on average less endowed with genes for high intelligence, while the upper classes are relatively more endowed with high-intelligence genes, thereby creating a **cognitive elite**.
　　4. The authors have been strongly criticized for their methodology and for largely ignoring the abundant research indicating that intelligence tests and standardized ability tests do not measure intelligence and ability as accurately for some groups as for others.

D. Tracking and Labeling Effects

1. Over one-half of elementary and secondary schools in the U.S. currently use some kind of **tracking,** or *ability grouping*, to separate students according to some measure of cognitive ability.
2. The basic idea behind tracking is that students will get a better education and be better prepared after high school if they are grouped according to ability.
3. Proponents of *detracking* argue that including students of varying cognitive abilities is more beneficial because students will learn from each other. They also believe that students in lower tracks get less teacher attention, thus, they learn less than when they are included in mixed groups.
4. One of the most consistent findings from research on tracking is that students in the higher tracks receive more positive effects, but lower track students suffer more negative effects, including being taught less.
5. Both high- and low- track students are subject to the **labeling effect,** whereby students are assigned to particular tracks and thereby labeled, whether or not the label accurately reflects the student's ability.

E. Teacher Expectancy Effect

The **teacher expectancy effect** refers to the influence of teacher expectations on a student's actual performance, independent of his or her actual ability. The expectations a teacher has for a student can dramatically influence how much the student will learn, as illustrated by Rosenthal and Jacobson's study of elementary schools. Through the **self-fulfilling prophecy**, merely applying a label has the effect of justifying that label.

F. Schooling and Gender

Girls and boys start out in school roughly equal in skills and confidence, but things change. Teachers hold different expectations about boys and girls in school, which affects students' actual performance. A comprehensive report prepared by the American Association of University Women (AAUW) summarized the findings of over 1,000 studies of gender and schooling.

1. Teachers pay less attention to girls and women than boys and men, particularly in math and science classes.
2. Women lag behind men in math and science ability and achievement scores.
3. Some standardized math and science tests still retain gender bias.
4. Standardized tests in math tend to under-predict women's actual grades in mathematics.
5. Teachers are more likely to rebuff Black girls and interact more with White girls.
6. Textbooks still tend to either ignore women or stereotype them.
7. As girls and boys approach adolescence, their self-esteem tends to drop, with the erosion of self-esteem occurring more quickly among girls than boys.

G. Stereotype Threat
1. A negative stereotype about one's self can affect one's own behavior. Research by Steele and associates indicates that Black students internalize the stereotype that Blacks have some inherent "deficiency" in math and verbal ability.

2. Internalization of such stereotypes leads to the **stereotype threat effect**. Consequently, when Black students and female students are told in advance that a test they will take is a "genuine" test of ability, they receive lower scores than do White males. When they are told nothing, they perform about the same as White male students.

IV. **HEALTH CARE IN THE UNITED STATES**

A. The Emergence of Modern Medicine

Citizens of the United are quite healthy in relation to the rest of the world; however, there are great discrepancies among Americans in terms of longevity, general health, and access to health care. By the start of the 19th century, the emergence of *germ theory* and other advances in biology and chemistry ignited a century of explosive growth in medical knowledge. The American Medical Association (AMA) was founded in 1847, and the social prestige of medicine, and its dominance over other forms of healing, greatly increased. The upper-class profession of medicine became identified with wealthy Whites, while more traditional folk practices and midwifery remained identified more with lower-class African Americans, Hispanics, and Native Americans − a trend that continues to this day.

B. Health, Diversity and Social Inequality

Prominent problem areas in the U.S. health care system include:
• **Unequal distribution of health care by race-ethnicity, social class, or gender.**
• **Unequal distribution of health care by region.**
• **Inadequate health education of inner-city and rural parents.**

The definitions of *sick* and *well* have varied greatly over time. For example, since the1950s, the United States has assigned positive value to being thin. The prevalence of eating disorders has increased in the American population, where **anorexia nervosa** particularly affects young, White women from upper and middle class families where parents place great pressure on their daughters to be high achievers. Men are not exempt from social pressure to achieve a particular body type, as evidenced by their increased use of *anabolic steroids*, which can have devastating consequences for men's health.

C. Race and Health Care
1. **Epidemiology** is the study of all the factors − biological, social, economic, and cultural − that are associated with disease. **Social epidemiology** is the study of the effects of social, cultural, temporal, and regional factors on health.
2. Race-ethnicity, class, gender, and age are among the more important factors affecting health in the United States. Although differences in culture, diet, and lifestyle account for some differences in health across racial groups, racial minorities do not receive equivalent medical treatment compared to Whites.
   a. African American women are more likely than Whites to suffer from cancer, heart disease, stroke, diabetes, and death during childbirth.
   b. Mortality rates are one and one-half times greater for Native Americans than for the general population.
   c. Hispanics have infant mortality rates similar to African Americans and Native Americans, and are even less likely than those groups to use available health services due to language barriers.

3. One of the challenges for the health care system is responding to the different cultural orientations of various groups in an increasingly diverse society. Developing the ability for greater cross-cultural administration of health care will likely continue to be a challenge in the future.

D. Social Class and Health Care
1. Social class has a pronounced effect on health and the availability of health services in the United States. The effects of social class are evident in the infant mortality and stillbirth rates as well as the distribution of diseases such as diabetes, tuberculosis, heart disease, cancer, and arthritis.
2. Although personal habits, such as smoking, are partly responsible for disparities in health, social circumstances are also influential. For example, stress due to financial difficulties, poor living conditions, elevated levels of pollution in low-income neighborhoods, and lack of access to health care facilities also contribute to the high rate of disease among lower-income people.
3. Sociologists have found that during interactions between health care providers and poor patients, especially those who are Black or Hispanic, the patients are more likely to be *infantilized* and to receive health counseling that is incorrect, incomplete, or delivered in language not likely to be understood by the patient.

E. Gender and Health Care
1. Even though women live longer on average than men, older women are more likely than older men to suffer from stress, hypertension, and chronic illness.
2. Some of the differences in health between men and women are due to environmental and life style differences.
3. The effects of employment on women's health can be positive or negative. For example, women are more likely to be "tokens" in the workplace, an experience that contributes to a greater incidence of depression and anxiety. However, housewives have higher rates of illness than women who work outside the home.

F. AIDS and Sexually Transmitted Diseases (STDs)
1. AIDS (Acquired Immune Deficiency Syndrome) is the term for a category of disorders that result from a breakdown of the body's immune system. Attitudes toward people suffering from sexually transmitted diseases, particularly AIDS, represent one of the strongest examples of how stigma operates in our society. **Stigma** occurs when an individual is socially devalued because of having some malady, illness or similar misfortune.
2. Death rates from AIDS present a picture similar to the incidence of AIDS. In 1995, the death rate was highest among African Americans, next highest among Hispanics, next among Whites, and lowest among Native Americans, Alaskan Natives, and Asians. Most of the negative stigma carried by AIDS victims is due to beliefs about who gets AIDS and how the HIV virus is transmitted.
3. A little over half of all cases of full-blown AIDS are the result of male-to-male sexual contact, followed by injecting drug use (24%). Women, who acquire HIV infection through injecting drug use or sexual contact with HIV-infected men, account for 19% of AIDS cases.

V. **THEORETICAL PERSPECTIVES ON HEALTH CARE**

A. The Functionalist View

1. Functionalist theory developed the notion of the **sick role**, defined as a pattern of expectations that society applies to someone who is ill. Functionalism asserts that any institution, group, or organization can be examined by looking at its positive and negative functions in society.

2. Positive functions, such as the prevention and treatment of disease, contribute to the harmony and stability of society, while negative functions contribute to social instability.

B. The Conflict Theory View

Conflict theory stresses that structural inequality is inherent in capitalist society, and this inequality is responsible for different groups' unequal access to medical care. Restricted access to health care is further exacerbated by the high costs of medical care; excessive bureaucratization, which leads to the alienation of patients; and prolonged waits in the emergency rooms of many urban hospitals in the United States.

C. Symbolic Interactionism

Symbolic interactionists assert that illness is partly socially constructed, resulting in the definitions of illness and wellness being culturally relative and time-dependent. The health care system itself has a socially constructed aspect. For example, one socially constructed problem is the tendency of medical practitioners to subject patients to *infantilization*, which refers to the treatment of adults as if they were children. Some medical schools now train their students to be more sensitive to their patients' feelings.

VI. **THE HEALTH CARE CRISIS IN THE UNITED STATES**

The cost of medical care in the United States is currently more than 14 percent of the gross national product, making health care the nation's third leading industry. The United States tops the list of all countries in per-person expenditures for health care, yet serious problems have placed the nation's health care system in crisis.

A. The Cost of Health Care
1. The central element in the U.S. health care payment system is the fee-for-service principle, whereby the patient is responsible for paying the fees charged by the physician or hospital.
2. Those with health insurance pass on their expenses to the insurance company, but there is a large proportion of uninsured, reaching 15% of the population in some areas of the U.S., particularly in the southern and southwestern states. The government is increasingly the ultimate payer of health care expenses.
3. Soaring costs of hospital care, the rise in fees for physician services and specialists, and the third party payment system all contribute to skyrocketing health care costs.
4. Two government programs provide guaranteed health service to particular populations. The **Medicare** program provides medical insurance covering primarily hospital costs for persons 65 and older. **Medicaid** provides health insurance for the poor, welfare recipients, and the disabled.

B. Medical Malpractice
1. There has been an increase in the number of patients who sue their physicians. The American public has traditionally accorded high social status and incomes to physicians, but the recent popularity of malpractice suits suggests that the public is beginning to question the privileged status of doctors.
2. Several specific reasons have been suggested for client revolts against the medical profession, including:

     a.   declining standards of care and a higher incidence of medical negligence;

     b.   the failure of doctors to establish rapport with patients, making patients less attached to their physicians and more likely to turn hostile; and

     c.   increasing animosity of patients who resent the high cost of medical care and the generous incomes of physicians.

   3. Doctors are increasingly practicing **defensive medicine**, which entails ordering expensive and excessively thorough tests at the least indication that something is wrong in order to protect themselves against potential malpractice suits.

C. <u>A Response to the Problem: HMOs</u>

   1. **Health Maintenance Organizations (HMOs)** are private clinical care organizations that provide medical services in exchange for a set membership fee; thus, they have direct responsibility and control over the costs incurred.

   2. Doctors in an HMO earn salaries rather than fees, and services are ultimately paid for by the membership fee that subscribers give to the HMO, thereby eliminating both the fee-for-service system and third-party insurers.

   3. Critics of HMOs argue that the system decreases the physician's right to determine treatments and limits the patient's right to choose a doctor.

D. <u>The Universal Health Care Debate</u>

The core of a universal health care plan is **managed care**. Under such a plan, everyone would belong to a complex of managed care organizations that would use their collective bargaining force to drive down health care costs. Individuals would still be free to retain their personal physicians, as long as their physicians met government-stipulated criteria.

INTERNET EXERCISES

1. Go to the homepage of the American Association of University Women website (www.aauw.org/) and click on AAUW Issue Advocacy and then Fact Sheets. Choose a topic that interests you, read the fact sheet, and write a brief assessment. On what specific research-based principles is the analysis or position based? Do you agree with the analysis or position? Why or why not?

2. National Center for Educational Statistics (www.ed.gov/NCES/) provides comprehensive statistics and research on all aspects of education in the United States, nationally and state-by-state. Explore the site, and then select NCES Fast Facts or Quick Tables and Figures. Select an issue that interests you and find the answer to one question.

3. The Centers for Disease Control and Prevention website (www.cdc.gov/) has extensive information and statistics on the full range of health-related issues from anthrax to youth violence. Use the search function to access reports and data on a topic that interests you, like HIV/AIDS, reproductive technology, aging, or obesity.

INFOTRAC EXERCISES

Conduct a keyword search using InfoTrac College Edition to extend the discussion in DOING SOCIOLOGICAL RESEARCH: *Beauty and Health* (p. 370-1).

1. **Keyword: Beauty myth.** This search leads to articles linking the beauty myth to appearance-based hiring and women in the workplace, feminist perspectives, related economic issues and coping.

2. **Keyword: Beauty and gender.** This search yields articles on idealized female body images, commerce, appearance, beauty therapy as emotional labor, inequality, women and eating disorders, dieting, and depression.

3. **Keyword: Anorexia.** This search yields a large number of articles that can be narrowed by searching only refereed articles. Topical links include therapy, healing, causes and cures, as well as adolescent males and females and older women.

PRACTICE TEST

Multiple Choice Questions

1. According to conflict theory, _____.
   a. schools equally prepare all children for participation in society by teaching them essential skills and knowledge
   b. schools are tools for discrimination through policies like tracking that perpetuate social inequalities
   c. the most important determinant of a child's success in school is how well the child gets along with his or her teachers
   d. schools today fulfill functions that were historically provided by other social institutions, such as the family

2. Which of the following is a latent function of the educational institution in the United States?
   a. Passing on cultural heritage, including ethics and norms of the society
   b. Training adolescents for jobs as adults by transmitting knowledge and skills
   c. Inculcating values that are important to the workplace, such as punctuality
   d. Reducing juvenile delinquency by keeping youth off the streets during the day

3. The extent to which a test score correlates with future college grades or some other criterion, such as likelihood of graduation, is called _____.
   a. general accuracy
   b. predictive validity
   c. cognitive reliability
   d. achievement potential

4. According to Herrnstein and Murray, the authors of *The Bell Curve*, differences in test performance between Black and White students result mainly from differences in _____.
   a. home environments, including the availability of books and educational toys
   b. nutritional deficiencies in infancy and early childhood
   c. the form of the test given to each group
   d. heredity or genetic makeup

5. Which of the following is a criticism of the results reported in *The Bell Curve*?
   a. The authors overemphasize environmental factors and dismiss the influence of heredity, or genetic makeup, on intelligence.
   b. The authors draw too many conclusions about between-group differences from within-group results.
   c. The authors heavily rely on the assertion that standardized tests are not accurate measures of intelligence for racial minority groups.
   d. all of these choices

6. Jamie was labeled "bright" and placed in the college preparatory track in high school, where she was encouraged and praised. She therefore enjoyed school and developed high educational and occupational aspirations. This reflects the _____.
   a. hidden curriculum
   b. detracking process
   c. self-fulfilling prophecy
   d. stereotype vulnerability

7. Based on a study commissioned by the American Association of University Women (AAUW), researchers have found that _____.
   a. in general, teachers pay less attention to male students than female students
   b. standardized math tests tend to over-predict women's actual grades in mathematics
   c. self-esteem erodes more quickly for girls than boys as they approach adolescence
   d. all of these choices

8. According to symbolic interaction theory _____.
   a. in terms of securing a high-paying job, the dollar value of a college degree has declined so much that getting a degree is no longer cost-effective
   b. standardized tests are an important mechanism for eliminating certain groups from competition for coveted spaces in universities
   c. one of the most important determinants of a child's success in school is what kind of perceptions and expectations his/her teachers have of and for him/her
   d. contemporary schools provide many functions for children and youth that used to be the responsibility of families and religious organizations

9. Black students perform less well than White students on math or verbal ability tests that they believe are "genuine" tests of their "true" ability; when they are told nothing, they perform about the same as White students. This phenomenon illustrates the _____.
   a. bell curve effect
   b. stereotype threat effect
   c. race factor
   d. labeling effect

10. Joe failed ninth grade. When his brother Frank entered ninth grade, he was assigned the same teacher as Joe. The teacher anticipated that Frank would be a poor student and invested little effort in helping Frank learn. At the end of the year, Frank nearly failed. This is an example of _____.
    a. student interaction effect
    b. teacher expectancy effect
    c. stereotype vulnerability
    d. educational deflation

11. Which of the following statements is **true** concerning the labeling effect?
    a. Once a student has been labeled as having a low ability level, that label can easily be altered if the student performs better on subsequent measures of his/her ability.
    b. Assigning labels to students is a fair, objective, and efficient mechanism for separating children into groups where they will be assigned material that is consistent with their ability levels.
    c. Once a label concerning ability is assigned to a student, it tends to stick, whether or not it is accurate.
    d. none of these choices

12. Which government program provides medical insurance to citizens who are 65 years of age and older to subsidize the cost when they are hospitalized?
    a. Social Security
    b. Eldercare
    c. Medicaid
    d. Medicare

13. Which theory argues that the inequality inherent in capitalist society is responsible for the unequal access various social groups have to medical care?
    a. symbolic interaction
    b. functionalist
    c. defensive
    d. conflict

14. In the United States, health care for most citizens is paid for through the fee-for-service system, which places primary responsibility for payment on _____.
    a. pharmaceutical companies
    b. government programs
    c. medical practitioners
    d. private citizens

15. Which of the following statements about medical malpractice in the United States is (are) **true**?
    a. The number of patients who sue their physicians is steadily increasing.
    b. The number of tests that doctors order for their patients is declining due to high costs.
    c. Lawyers are reluctant to accept most malpractice cases because they do not believe that medical errors occur that often.
    d. all of these choices

16. Which of the following statements about universal health care proposals in the United States is (are) accurate?
    a. The core of proposed universal health care plans is the principle of managed care.
    b. Under a universal health care plan, individuals would not be able to retain their personal physicians.
    c. Universal health care plans have no mechanism for containing costs.
    d. all of these choices

17. The eating disorder anorexia nervosa is most likely to afflict _____.
    a. young, White women, from well-to-do families, most often two-parent families
    b. African American women and Latinas
    c. young athletic men
    d. young women from single-parent families and lesbians

18. Which of the following statements accurately reflect the findings of social epidemiology in the U.S.?
    a. Among Native Americans under the age of 45, the death rate is three times that of Whites.
    b. African American women are three times more likely to die while pregnant than White women.
    c. Hispanics are less likely than Whites to have a regular source of medical care.
    d. all of these choices

19. Which of the following statements about AIDS in the U.S. is accurate?
    a. A little more than one-third of new AIDS cases are the result of male-to-male sexual contact.
    b. The AIDS rate among Whites is much higher than the rate among African Americans.
    c. Women now account for almost half of AIDS cases.
    d. Most AIDS cases today are the result of intravenous drug use.

20. Which of the following is a major problem affecting the U.S. health care system?
    a. Unequal distribution of health care by race-ethnicity, social class, or gender
    b. Unequal distribution of health care by region
    c. Inadequate health education of inner-city and rural parents
    d. all of these choices

True-False Questions

1. Although the average income for women without a high school diploma is less than the average income for men with the same education level, women and men with college degrees earn about the same average income.

   TRUE or FALSE

2. Among White people of the upper classes, education is more important than social class origin in determining occupation and income.

TRUE or FALSE

3. As family income increases, students' Scholastic Aptitude Test (SAT) scores dramatically increase.

TRUE or FALSE

4. The SAT test in the U.S. has an effect similar to a test administered in England, of diminishing the chances for lower-class and minority students to be admitted to the best colleges and universities.

TRUE or FALSE

5. White students are more likely than Black or Hispanic students to be placed in high ability tracks in school, even when the students receive the same test scores.

TRUE or FALSE

6. Medicare is a form of medical insurance provided by the government to poor people and people with disabilities.

TRUE or FALSE

7. Full-time housewives have lower rates of illness than do women who work outside the home.

TRUE or FALSE

8. The occurrence of breast cancer is lower for Black women than White women, but the mortality rate from breast cancer is considerably higher for Black women than White women.

TRUE or FALSE

9. Functionalist theory argues that the health care system contributes to the stability of society by preventing and treating disease in the population.

TRUE or FALSE

10. Unlike other minorities in the United States, Native Americans have access to a superior health care system.

TRUE or FALSE

Fill-in-the-Blank Questions

1. Exams that are given to large populations and scored with respect to population averages are

   _____ tests.

2. People are especially likely to assign a _____ to people who are ill from sexually transmitted diseases such as AIDS.

3. Medical practitioners who treat adult patients like children by speaking to them in a condescending or patronizing manner are subjecting their patients to _____ .

4. Doctor who order excessively thorough tests, X-rays, and so on to protect themselves against potential malpractice suits are practicing _____ .

5. The insistence upon educational degrees or diplomas even if they bear little relationship to the intended job is called _____ .

Essay Questions

1. Compare and contrast the functionalist, conflict, and symbolic interaction perspectives on the role of education in society.

2. Does schooling matter in the United States?

3. Discuss why Herrnstein and Murray's book, *The Bell Curve*, has generated so much controversy in the United States.

4. Explain why there are dramatic difference in life expectancy among different racial-ethnic and social class groups in the United States.

5. Identify three of the most pressing problems in the United States health care system today, and suggest at least one solution for addressing each of those problems.

Solutions

PRACTICE TEST

Multiple Choice Questions

1. B, 352-5 Conflict theory views education as reinforcing and perpetuating inequalities in the society at large through processes such as tracking. Functionalist theory focuses on the functions provided by social institutions, noting that schools prepare all children equally for participation in society. Symbolic interaction focuses on interactions between teachers and students [Table 13.1].

2. D, 354 Social control is a latent function of education. A perceived benefit of compulsory education is that it keeps young people out of trouble.

3. B, 359 The extent to which a test accurately predicts future college grades, likelihood of graduation, or some other criterion is predictive validity.

4. D, 360 In *The Bell Curve*, Herrnstein and Murray argue that heredity, or genetic makeup, determines 70 percent of intelligence. They assert that the upper class constitutes a genetic cognitive elite in America.

5. B, 360-1 *The Bell Curve* authors have been criticized for drawing too many conclusions about between-group differences from within-group results. The authors dismissed previous research that indicated environment significantly contributes to differences in test performance, and that standardized tests do not accurately measure intelligence for all racial-ethnic or social class groups.

6. C, 362 The self-fulfilling prophecy is a powerful mechanism whereby labeling a student has the effect of justifying the label and producing the expected behavior.

7. C, 363-4 Both girls and boys experience a drop in self-esteem as they approach adolescence, but the erosion of self-esteem occurs more quickly for girls. In schools, teachers generally pay more attention to male students. Math tests tend to under-predict girls' actual grades in mathematics.

8. C, 354 Symbolic interaction theory focuses on the nature and quality of interaction between people in social settings, such as students and teachers in schools. This theory recognizes that how a teacher perceives and treats students, depending on their gender, race, and class, affects their school performance.

9. B, 364 Claude M. Steele and associates studied the impact of internalized stereotypes on women and Black college students in high stakes tests, which students believed reflected their "true" ability. They called this the stereotype threat effect.

10. B, 362 The teacher expectancy effect refers to the process whereby the teacher's expectations affect the student's actual performance, independent of the student's actual ability.

11. C, 361-2 Both high- and low-track students are subject to the labeling effect. Once a label has been assigned, it tends to stick, whether or not it is accurate. These labels are not necessarily based on fair, objective, or relevant criteria for determining a student's ability.

12. D, 375 Started in 1965, the Medicare program is a form of health insurance provided by the government to people age 65 and older to subsidize hospitalization costs. Medicaid is a government program that provides medical insurance to poor people and people with disabilities.

13. D, 354 Conflict theory views differential access to health care as the result of the economic inequality inherent in capitalist society.

14. D, 374 The fee-for-service system of medical care places responsibility for payment on private citizens. The government provides medical insurance to some vulnerable groups, such as the elderly, veterans, poor people, and people with disabilities.

15. A, 375-6 The number of patients who sue their physicians is increasing, and the trend toward filing malpractice suits is supported by lawyers who argue that medical errors and negligence have increased. In response, physicians are practicing defensive medicine, whereby they order more tests for patients, despite the cost.

16. A, 376-7 Universal health care proposals are based on the principle of managed care, which creates a collective bargaining force that keeps down costs, and

under which individuals could retain their personal physicians.

17. A, 366 Anorexia nervosa is most likely to affect young, White4 women, from well-to-do families, most often two-parent families. It is less likely to affect African American or Latina women, lesbians, or males.

18. D, 367 Minorities experience lower life expectancy and healthcare access than Whites.

19. A, 371 About one-third of AIDS cases result from male-to-male sexual contact. One-third of

new AIDS cases are women who contract AIDS through heterosexual contact. AIDS disproportionately hits inner-city minority communities.

20. D, 365 Unequal distribution of health care by race-ethnicity, social class, gender, and region all affect the U.S. healthcare system. Inner-city and rural parents are affected by inadequate health education.

True-False Questions

1. F, 356 Table 13.2 indicates that at all education levels, women's average income is lower than men's average income in the United States.

2. F, 356 Among White people of the upper classes, social class origin is more important than education in determining occupation and income.

3. T, 357 Table 13.3 indicates that there is a dramatic increase in students' SAT scores as family income increases.

4. T, 356-7 England's Eleven-Plus test has an effect similar to the U.S. SAT of diminishing the chances for lower-class and

minority students to get into the best colleges.

5. T, 361-2 Race and ethnicity continue to influence tracking decisions in schools.

6. F, 375 Medicaid is a government program that provides medical insurance to poor people and people with disabilities. Medicare is a government program that provides insurance to elderly people to pay for hospitalization.

7. F, 369-70 Full time housewives have higher rates of illness than do employed women.

8. T, 367 White women are more likely than Black women to

develop breast cancer, but when the cancer is developed, Black women are more likely to die of the disease.

9. T, 372 Functionalist theory views the health care system as contributing to the stability of society by preventing and treating illness.

10. F, 377 The mortality rate for Native Americans is one and one-half times that of the general population. The limited delivery of health care services to Native American people is a particularly serious problem in the U.S.

Fill-in-the-Blank Questions

1. standardized, 358

2. stigma, 370

3. infantilization, 373

4. defensive medicine, 376

5. credentialism, 354

Essay Questions

1. Sociological Theories of Education. The three major theoretical perspectives are applied to education on p. 352-355. Student responses should demonstrate an understanding of the theories and be able to apply them to education and the role of education in society.

2. Does Schooling Matter? This question is addressed on p. 354-357. Student responses should demonstrate an understanding of the relationship between education, occupation, income, social mobility, and social class background.

3. The Bell Curve. The Bell Curve is discussed on p. 360-361. Student responses should demonstrate awareness of the arguments put forward by Herrnstein and Murray, and the critiques of their position.

4. Life Expectancy and Inequalities. The impact of race/ethnicity and social class on health and life expectancy are discussed on p. 365-368. Student responses should demonstrate an understanding of the differential impact of environment, healthcare access, and healthcare education.

5. Problems with the U.S. Health Care System. The three major problems in the U.S. health care system are discussed on p. 365. Student responses should demonstrate awareness of these problems and their relationship to race-ethnicity, region, and social class.

POLITICS AND THE ECONOMY

BRIEF CHAPTER OUTLINE

Defining the State

The Institutions of the State

The State and Social Order

Revolution

Global Interdependence and the State

Power and Authority

Types of Authority

The Growth of Bureaucratic Government

Theories of Power

The Pluralist Model

The Power Elite Model

The Autonomous State Model

Feminist Theories of the State

Government: Power and Politics in a Diverse Society

Diverse Patterns of Political Participation

Political Power: Who's in Charge?

Women and Minorities in Government

The Military

Economy and Society

The Industrial Revolution

Comparing Economic Systems

The Changing Global Economy

A More Diverse Workplace

CHAPTER FOCUS

This chapter examines the social institutions of politics and the economy. The major political concepts of state, revolution, and authority, as well as the major theories of power are examined. Patterns of political participation in the U.S. and issues related to the U.S. military are also explored. The chapter examines the major economic systems, the global economy, the effects of deindustrialization and technological change on work and the economy, characteristics of the labor force and issues related to the workplace.

QUESTIONS TO GUIDE YOUR READING

1. What is the state, what are its main institutions, and how does the state maintain social order?

2. What are the characteristics, conditions, causes, and possible outcomes of revolutions?

3. What are the major types of authority, and what is behind the growth of bureaucratic government?

4. What are the strengths and weaknesses of the major theories of power: pluralist, power elite, autonomous state, and feminist?

5. What explains the diverse patterns of political participation in the U.S.?

6. What is the structure and composition of the U.S. military today, and how do race and the inclusion of women, gays, and lesbians influence the military?

7. What are the major economic systems, and what is the structure of the global economy?

8. What are the effects of deindustrialization and technological changes on the economy and work?

9. How is work defined sociologically, what is the division of labor in the U.S., and how do the three major theoretical perspectives approach work?

10. What are the characteristics of the labor force, including unemployment patterns, dual labor market, and occupational distribution by race, class, and gender?

11. What are the dynamics of power in the workplace with respect to the glass ceiling, sexual harassment, gay and lesbian experience, and disabilities?

KEY TERMS

(defined at page number shown and in glossary)

CHAPTER OUTLINE

I. **DEFINING THE STATE**

The **state** is an abstract concept that includes the institutions that represent official power in society, such as the government and its legal system, the police, and the military. The state exists to regulate social order and has a central role in determining the rights and privileges of different groups. Sociological analyses of the state focus on several important issues such as the relationship between the state and inequality in society, and the connections between the state and other social institutions, including religion and the family.

A. The Institutions of the State
   1. The government creates laws and procedures that govern society.
   2. The court system punishes wrongdoers and adjudicates disputes, and the prison system punishes those who have violated the law.
   3. The law is a type of formal social control that outlines what is permissible and forbidden in a society.
   4. The police are responsible for enforcing the law and maintaining public order at the local level.

B. The State and Social Order
   1. Through its power to regulate the media, the state influences public opinion. The state can also direct public opinion through censorship or by circulating **propaganda** or information disseminated by a group or organization that is intended to justify its power.
   2. The state's role in maintaining public order is apparent in how it manages dissent. For example, states may use options such as surveillance, imprisonment, and military force to respond to protests, as seen in the Civil Rights Movement.
   3. Even in a democratic state like the United States, the state typically protects the interests of the groups with the most power, leaving the least powerful groups in society more vulnerable to oppressive state action.

C. Global Interdependence and the State

   On an international level, there are increasingly strong ties between the state and the global economy. Globalization has profound effects on the relationships of states to each other. For example, the World Trade Organization was created in 1984 to monitor and resolve disputes between nations.

II. POWER AND AUTHORITY

Power is the ability of one person or group to exercise influence and control over others. Sociologists are most interested in how power is structured at the societal level, who has it, how it is used, and how it is built into institutionalized structures. **Authority** is power that is perceived by others to be legitimate. Authority emerges not only from the exercise of power, but also from the constituents' belief that the power is legitimate. Those who accept the status quo as a legitimate system of authority perceive the guardians of the law to be exercising *legitimate power*. *Coercive power* is achieved through force, often against the will of the targeted people.

A. Types of Authority

   Max Weber identified three types of authority in society.

   1. **Traditional authority** stems from the long-established patterns that give certain people or groups legitimate power in society, such as a monarchy.

2. **Charismatic authority** is derived from the personal appeal of a leader, who is often believed to have special gifts, even magical powers, which inspire devotion and obedience.

3. **Rational-legal authority**, the most common form of authority in the contemporary United States, stems from rules and regulations, typically written down as laws, procedures, or codes of conduct. Under rational-legal authority, authority is gained through having been elected or appointed in accordance with society's rules.

B. The Growth of Bureaucratic Government
    1. According to Weber, rational-legal authority inevitably leads to the formation of **bureaucracy**, a type of formal organization characterized by an authority hierarchy, a clear division of labor, explicit rules, and impersonality.
    2. Bureaucracy, the modern system of administration, is supposed to be a highly efficient mode of organization, although the reality is often much different.
        a. Bureaucracies tend to proliferate rules, often to the point that the organization becomes ensnared in its own bureaucratic requirements, and the actual work of the system bogs down.
        b. Within bureaucracies, personal temperament and individual discretion are not supposed to influence the application of rules; however, bureaucratic workers frequently exercise discretion in applying rules and procedures, and people may receive widely different treatment based on their race, gender, age, or other characteristics.

III. **THEORIES OF POWER**

Table 14.1 summarizes the four different models sociologists have developed to explain how power is exercised in society.

A. The Pluralist Model
    1. The **pluralist model** interprets power in society as coming from the representation of diverse interests of different groups in society.
        a. This model assumes that in democratic societies, the system of government works to balance the different interests of groups in society.
        b. An **interest group** can be any constituency in society organized to promote its own agenda, including large, nationally-based groups such as the American Association of Retired Persons (AARP); groups organized around professional interests, such as the American Medical Association (AMA); or groups that concentrate on a single political or social goal, such as Greenpeace.
        c. According to the pluralist model, interest groups achieve power and influence through their organized mobilization of concerned people.
    2. The pluralist model, which views the state as representative of the whole society, has its origins in functionalist theory.
        a. Power is broadly diffused across the public, and any group that wants to effect a change or express their opinion need only mobilize to do so.
        b. Because the state is benign and representative of the whole society, different interest groups compete for government attention with equality of political opportunity.

      c.   This model helps explain the importance of **political action committees (PACs),** groups of people who organize to support candidates they feel will represent their views.

B. <u>The Power Elite Model</u>

    1. The **power elite model** originated in the work of **Karl Marx** (1818-1883), who argued that the dominant group controls the main institutions in society. The state is simply an instrument by which the ruling class exercises its power, rather than a representative, rational institution.

    2. **C. Wright Mills** (1956) elaborated on Marx's work in his analysis of the power elite, arguing that the true power structure consists of people well-positioned in three areas: the economy, the government, and the military.

        a.   The power elite model posits a strong link between government and business, a view that is supported by the role of military spending as a principal component of economic affairs in the United States.

        b.   This model emphasizes how power overlaps between influential groups. **Interlocking directorates** are organizational linkages created when the same people sit on the boards of directors of several corporations.

C. <u>The Autonomous State Model</u>

    1. The **autonomous state model** interprets the state as its own major constituent; thus, the state seeks to promote its own interests, independent of other interests and the public that it allegedly serves.

    2. This theory emphasizes that states tend to grow over time, leading to a paralyzed government in conflicts between revenue-seeking state bureaucrats and those who must fund them. This can in turn lead to a revolt against the state, as illustrated by recent tax revolts in the U.S.

D. <u>Feminist Theories of the State</u>

    1. Feminist theory begins with the premise that an understanding of power cannot be achieved without a strong analysis of gender.

    2. The feminist model concludes that the state is devoted primarily to men's interests, and the actions of the state tend to support gender inequality.

    3. Groups that exercise state power, such as Congress and the military, are predominantly male, and these organizations are structured by values that are culturally masculine.

IV. GOVERNMENT: POWER AND POLITICS IN A DIVERSE SOCIETY

The **government**, one of several institutions that make up the state, includes those state institutions that represent the population and make rules that govern the society. Although the United States is a **democracy**, or a system of government based on the principle of representing all people through the right to vote, the actual composition of the government is not representative of the nation's population.

A. <u>Diverse Patterns of Political Participation</u>

    1. Among democratic nations, the United States has one of the lowest voter turnouts, with 50 percent or less of the eligible voters typically participating in national elections; in the 2004 election, only 61% of eligible voters voted [Figure 14.1].

2. In sociological terms, age, income, and education are the strongest predictors of whether someone will vote. The group most likely to vote is older, better educated, and financially better off than the average American citizen.

3. There is significant variation in voting patterns by race, with African Americans being less likely to vote than Whites, which reflects the disproportionate numbers of African Americans in the poor and working classes.

4. Race, ethnicity, and gender also influence for whom people vote; for example, female, African American, and Latino voters are more likely than male and White voters to vote as Democrats.

5. **Gender gap** refers to the differences between men and women in political attitudes and behaviors.

B. Political Power: Who's in Charge?

Most of the 535 members of Congress are well-educated White men from upper middle class or upper class backgrounds who have Anglo-Saxon Protestant heritage. Simply getting into politics requires a substantial investment of money, and many of the members of Congress are millionaires. Surveys indicate that most American citizens have little confidence in the political process, and few trust government officials.

C. Women and Minorities in Government

Whereas 51 percent of the population of the United States is female, there are currently only 14 women United States Senators. African Americans, Hispanics, Asian Americans, and Native Americans are vastly underrepresented at both the state and the federal levels of government [Figure 14.2].

D. The Military

The military is among the most powerful and influential social institutions in all societies. In the United States, it is the single largest employer, and a majority of Americans report having a lot of confidence in the military. *Militarism*, or the pervasive influence of military goals and values throughout the culture, is evident in the toys, movies, and fashions of the nation.

1. ***The Military as a Social Institution***. Institutions are stable systems of norms and values that fill certain functions in society. The military's function is to defend the nation against external, and sometimes internal, threats.
    a. The military hierarchy is extremely formalized, which is reflected in the practice of explicitly labeling its members with rank, and conformity is highly valued, as seen in solders' identical uniforms and haircuts.
    b. A striking feature of the military as a social institution is the diverse representation of racial minority groups and women within the armed forces, particularly since the elimination of the all-male draft in 1973 and the initiation of an all-volunteer force in the U.S. armed services.

2. ***Race and the Military***. African Americans have served in the military for almost as long as the armed services have been in existence, although the armed forces were officially segregated by race until 1948.
    a. Black Americans are the most over-represented group relative to their proportion in the civilian population, although there are still few people of color at the highest ranks.

      b. For groups with limited opportunities in civilian society, joining the military seems to promise an educational and economic boost. In fact, for both Blacks and Whites, serving in the military leads to higher earnings relative to one's non-military peers.

  3. ***Women in the Military***. There have been profound changes in the military since military academies began accepting women in 1976; however, there is still resistance to the full inclusion of women in the military, particularly in combat.

      a. In 1996, the Supreme Court ruled that women can not be excluded from state-supported military academies such as the Citadel and VMI.

      b. Gender relations in the military extend beyond the women who serve. For example, the experience of military wives is greatly affected by the frequent relocation associated with their husbands' employment.

  4. ***Gays and Lesbians in the Military***. According to the 1992 "*don't ask, don't tell*" policy, recruiting officers cannot ask recruits about sexual preference, and individuals who keep their sexual preference a private matter shall not be discriminated against. This controversial policy reveals continuing prejudice and discrimination against lesbian women and gay men. The longstanding policy against gays and lesbians in the military has kept homosexuality in the military hidden, but not nonexistent, while homophobia, disparaging comments and harassment toward gay service personnel persist.

V. ECONOMY AND SOCIETY

The economy is the system by which goods and services are produced, distributed, and consumed in a society. Sociologists understand work in the context of this social institution.

A. The Industrial Revolution

  1. The Industrial Revolution is giving way to the growth of post-industrial societies, which reflects a development in the economic system that has far-reaching consequences for how society is organized.

  2. The development of agricultural societies followed the introduction and use of technologies that enabled people to increase the production of food from simple hunting and gathering techniques to more large-scale production.

  3. Industrialization was probably the most significant historical development to affect the social organization of work.

  4. Although the United States is largely industrial, it is being quickly transformed into a **postindustrial society**, which is mainly organized around the provision of services such as banking, retail sales, and health care.

B. Comparing Economic Systems

The three major kinds of economic systems found in the world today are capitalism, socialism, and communism. These are ideal types, because many societies have a mix of economic systems.

  1. **Capitalism** is an economic system based on private ownership of the means of production and the principles of private property, market competition, and the pursuit of profit. The United States is a capitalist society.

2. **Socialism** is an economic institution characterized by state ownership and management of basic industries, where the means of production are the property of the state. The People's Republic of China was formerly a socialist society, but is now developing a mix of socialist and capitalist principles.

3. **Communism** is sometimes described as socialism in its purest form. In pure communism, the state is the sole owner of the systems of production. Nineteenth century communist theorists believed class divisions and private property would be eliminated when capitalism was overthrown during a workers revolution, but history has not borne out this prediction.

VI. THE CHANGING GLOBAL ECONOMY

The concept of a **global economy** acknowledges that all dimensions of the economy now cross national borders, including investment, production, management, markets, labor, information, and technology. The global economy now links the lives of millions of Americans to the experiences of other people throughout the world, primarily through **multinational corporations** that have become increasingly powerful. In the global economy, the most developed countries control research and management, while assembly line work is performed in nations with less privileged positions in the global economy. The relocation of manufacturing to places where labor is cheap and management is strong has led to the emergence of the *global assembly line*, a new international division of labor. Women and children in poor, underdeveloped nations now assemble most toys and other goods that are sold in the rich nations. In the United States, the development of a global economy has created anxieties about foreign workers, particularly among the working class, who are developing **xenophobia**, or the fear and hatred of foreigners. The development of a global economy is part of a broad process known as **economic restructuring**, which refers to transformations in the basic structure of work that are permanently altering the workplace.

A. A More Diverse Workplace

Economic restructuring has led to a more diverse workplace, including more women, racial minorities, and older workers. Most of the growth is in service industries, where better jobs require education and training. Manufacturing jobs are declining, as are entry-level corporate jobs. Many college graduates are underemployed, but still have higher earnings than those with less education.

B. Deindustrialization

1. **Deindustrialization** refers to the transition from a predominantly goods-producing economy to one based on the provision of services.
2. This process is most easily observed by looking at the number of jobs in the manufacturing sector of the United States economy since World War II, when the majority of workers were employed in manufacturing-based jobs. Today, at least 70 percent of workers are employed in the service sector, including childcare, food preparation, banking, and clerical work.
3. Among the areas that have been hardest hit by deindustrialization are communities that were heavily dependent on a single industry, including steel towns and automobile-manufacturing cities, where there are high rates of **job displacement** and few prospects for economic recovery.

C. Technological Change

1. Rapidly changing and developing technologies, such as the invention of the semiconductor, are bringing major changes in work, including how it is organized, who does it, and how much it pays.

2. Increasing reliance on the rapid transmission of electronic data has produced *electronic sweatshops*, a term referring to the back offices found in many industries where workers at computer terminals process hundreds of transactions per day.

3. **Automation**, the process by which human labor is replaced by machines, eliminates many repetitive tasks and facilitates rapid communication; however, it may also make workers subservient to machines.

4. *Deskilling*, a side effect of automation and technological progress in which the level of skill required to perform certain jobs declines over time, has several consequences.
   a. Employees are paid less and have less control over the tasks they do.
   b. Jobs require less mental labor and become routine and boring.
   c. There is greater polarization of the work force.

5. Along with deskilling, there has been an increasing reliance on temporary or **contingent workers**, those whose employment is contracted as needed. Although these jobs offer workers flexibility and autonomy, contingent workers are paid less and are less likely to receive benefits than those who hold regular jobs.

VII. THEORETICAL PERSPECTIVES ON WORK

A. Defining Work

1. Sociologists define **work** as productive human activity that creates something of value, either goods or services, whether or not a person gets paid for their labor.

2. Housework is not included in the official measure of productivity that economists use to indicate the work output of the United States.

3. Some sociologists argue that referring to only physical and mental labor as work is too narrow a definition. Arlie Hochschild introduced the concept of **emotional labor** to refer to another form of work that is common in a service-based economy, in which workers try to produce a certain feeling or desired state of mind in a consumer.

B. The Division of Labor

The division of labor is the systematic interrelatedness of different tasks that develops in complex societies. In the United States, the division of labor is most affected by gender, race-ethnicity, social class, and age.

1. The *class division of labor* can be observed by examining the work done by those with different educational backgrounds, because education is a fairly reliable indicator of class.

2. The *gender division of labor* refers to the different work that women and men do in society. There is a cultural tendency toward placing more value on men's than women's work.

3. The *racial division of labor* is seen in the pattern of people from racial and ethnic minority groups disproportionately working in the lowest paid, least prestigious, and most arduous jobs.

C. Functionalism, Conflict Theory, and Symbolic Interaction

1. **Conflict theorists** view the transformations taking place in the workplace as resulting from tensions inherent in the social system. These tensions develop because groups with unequal power are competing for economic resources.

2. **Functionalist theorists** interpret work and the economy as functional for society. When the society changes too rapidly, due to technological developments and globalization, for example,

institutions may experience social disorganization, and individuals may experience **alienation**, or feelings of powerlessness and separation from society.

3. *Symbolic interaction* is less focused on the workings of the whole society; but this theory would be useful for studying the meaning of work to those who do it, as well as how interaction in the workplace supports or undermines social bonds between people.

VIII. **CHARACTERISTICS OF THE LABOR FORCE**

Data on the characteristics of the American labor force are typically drawn from the official statistics reported by the United States Department of Labor. In 2003, 146 million people, or 66 percent of the working-age population, were in the labor force.

A. Who Works?
   1. Employment patterns vary considerably for different groups in the population. For example, Hispanic men are most likely, and Hispanic women are least likely, to be employed [Figure 14.3].
   2. One of the most dramatic changes in the labor force since World War II has been the number of women employed, which increased from 35 to 60 percent since 1948. This has resulted in greater demands on workplaces and society to provided childcare facilities and more progressive family leave policies.

B. Unemployment and Joblessness
   1. The U.S. Department of Labor regularly reports data on the **unemployment rate** or the percentage of people not working but officially defined as actively looking for work. Unemployment statistics are most likely to undercount the people for whom unemployment runs the highest, including the youngest and oldest workers, women, and members of racial minority groups.
   2. Migrant workers and other mobile populations are undercount, workers on strike are counted as employed, and official unemployment rates take no notice of **underemployment**.
   3. Native Americans, followed by African Americans and Puerto Ricans, are most likely to be unemployed.
   4. While popular explanations often attribute unemployment to the failings of individual workers, sociologists explain unemployment by examining structural problems in the economy, such as deindustrialization and discriminatory employment practices.
   5. Women are more likely than White men to be displaced from jobs. The employment of women and minorities does not cause White male unemployment, primarily because of race and gender segregation in the labor force.
   6. The contemporary labor force is being shaped by the employment of recent immigrants, which include both the most and the least educated segments of the population.

C. Diversity in the American Occupational System
   1. *The Dual (Segmented) Labor Market.* As a branch of conflict theory, **dual labor market theory,** views the labor market as composed of two major segments: the *primary labor market* and the *secondary labor market* [Table 14.4].
      a. The *primary labor market* offers jobs with relatively high wages, benefits, stability, good working conditions, opportunities for promotion, job protection and due process

for workers, which means that workers are treated according to established rules and procedures that are allegedly fairly administered.

 b. The *secondary labor market* is characterized by low wages, few benefits, high turnover, poor working conditions, little job protection or opportunity for advancement, and the arbitrary treatment of workers. Women and racial-ethnic minorities are the groups most likely to be employed in the secondary labor market.

 c. In addition to these two major segments, there is the *underground economy*, which includes both legal and illegal, unreported work.

2. ***Occupational Distribution.*** Workers are dispersed throughout the occupational system in patterns that vary greatly by race-ethnicity, class, and gender, revealing a pattern of **occupational segregation**.

 a. Work in the United States is classified into six broad categories: managerial and professional; technical, sales, and administrative support; service; precision production, craft, and repair; operators, fabricators, and laborers; and farming, forestry, and fishing.

 b. Several changes in occupational distribution have occurred, including: a decline in the number of Black women in private domestic work, an increase in the number of racial minorities in professional-managerial work (such as business, law, and municipal management), and an increase in the number of women employed in working-class jobs traditionally held only by men (such as precision, craft, and repair work).

3. ***Occupational Prestige and Earnings.*** There is a strong correlation between *occupational prestige* and the race and gender composition of workers in given jobs. Functionalist and conflict theorists explain this correlation differently.

 a. To functionalists, differential wages are a source of motivation for workers, and higher wages reflect the most valued characteristics that workers bring to jobs.

 b. From a conflict perspective, wage inequality is one way that the systems of race, class, and gender inequality are maintained, because some of the most essential jobs are the most devalued and poorly rewarded.

4. ***Work and Immigration.*** The proportion of professionals and technicians among legal immigrants exceeds the proportion of professionals in the labor force as a whole. Even illegal immigrants, who are mostly working class, have higher educational and occupational skill than the typical worker in their homeland. Yet immigrant workers tend to start at the bottom of the occupational ladder.

IX. POWER IN THE WORKPLACE

Obvious, tangible factors such as level of pay and benefits influence workers' degree of satisfaction with their jobs. Other factors include the value accorded to one's job and opportunity for advancement. The **glass ceiling** refers to the limits that women and racial-ethnic minorities experience in job mobility.

A. Sexual Harassment
1. **Sexual harassment** is defined as unwanted physical or verbal sexual behavior that occurs in the context of a relationship of unequal power and that is experienced as a threat to the victim's job or educational activities.
2. The law recognizes two forms of sexual harassment.

a. *Quid pro quo sexual harassment* forces sexual compliance in exchange for an employment or educational benefit.

b. The other form of sexual harassment is the creation of a *hostile working environment*, in which unwanted sexual behaviors are a continuing condition of work, including touching, teasing, and/or sexual comments.

3. Sexual harassment was first made illegal by Title VII of the 1964 Civil Rights Act, which identifies it as a form of employment discrimination. The Supreme Court upheld this principle in *Meritor Savings Bank v. Vinson* in 1986.

4. Sexual harassment tends to be underreported, but surveys indicate that half of all employed women experience this form of discrimination at some time.

B. <u>Gays and Lesbians in the Workplace</u>

As gays and lesbians are more open about their sexuality, more attention has been given to sexual identity and workplace experiences. Public opinion polls indicate that the majority of people are now accepting of gays and lesbians in the workplace, except as teachers and clergy. Any negative experience in the workplace can affect self-esteem, productivity, and economic well being. Many gays and lesbians fear that they will suffer adverse career consequences if their coworkers know that they are homosexual.

C. <u>Disability and Work</u>

Sociologist Irving Zola was one of the first to suggest that people with disabilities face issues similar to minority groups. This approach emphasizes the group rights of people with disabilities, who are protected from discrimination in education, employment, and public facilities (such as transportation) under the Americans with Disabilities Act, adopted by Congress in 1990.

INTERNET EXERCISES

1. The Federal Election Commission. (www.fec.gov/) documents campaign financing and other issues of public concern related to elections. Use the site to "follow the money" to track the disbursements and receipts from PACs and individuals to specific candidates or parties.

2. The National Institute for Occupational Safety and Health (NIOSH) (www.cdc.gov/niosh/homepage. html) has basic information on employee safety and health as well as highly technical studies, access to technical assistance, strategic plans, and other research. Use the site to investigate an issue of interest to you, like mining safety or health hazards in particular occupations, occupational violence, or women's work health.

3. The U.S. Department of Labor website (www.dol.gov/) provides a wealth of information, and is continually updated with new reports and data. Investigate a particular that interests you for a short report, or as a component in a larger research project. You can also access the Occupational Outlook Handbook, with which you could investigate the employment prospects in a field of interest to you.

INFOTRAC EXERCISES

Conduct a keyword search using InfoTrac College Edition to extend the discussion in DOING SOCIOLOGICAL RESEARCH: *All in the Family: Children of Immigrants and Family Businesses* (p. 402-3).

1. Keyword: Immigrant children. This search yields articles on the need for health insurance and medical screening, children mistreated in detention centers, education, and other issues.

2. Keyword: Immigrant labor. This search leads to articles on labor reform, legal status, US employers, benefits and drawbacks to the economy, and other issues.

3. Keyword: Immigrant worker. This search leads to articles on fatalities, worker rights, protections, legislation, and the impact of immigrant workers on the economy.

PRACTICE TEST

Multiple Choice Questions

1. Which of the following state institutions is responsible for maintaining public order and enforcing laws at the local level?
   a. government
   b. military
   c. prisons
   d. police

2. The state institution that creates laws and procedures to establish order in society is the _____.
   a. government
   b. military
   c. prisons
   d. police

3. According to Max Weber, which type of authority inevitably leads to the formation of bureaucracies?
   a. global
   b. traditional
   c. charismatic
   d. rational-legal

4. A monarchy derives its legitimate right to rule from which type of authority?
   a. global
   b. traditional
   c. charismatic
   d. rational-legal

5. Which sociological model of power interprets power in society as derived from the representation of the diverse interests of different groups in society?
   a. pluralist
   b. feminist
   c. power elite
   d. autonomous state

6. Feminist theories of the state assert that:
   a. power is derived from the organized activities of multiple public interest groups, all of which have equal opportunities to influence the political system.
   b. the state is fundamentally feminine in its values because decisions are based on the principle of caring for the nation's citizens.
   c. social conflict derives from the domination of the upper class elites over other groups with fewer economic resources.
   d. the state is fundamentally masculine in its values, and its actions tend to support gender inequality in society.

7. The landmark U.S. Supreme Court case, *United States v. Virginia* (1996), declared that _____.
   a. gay and lesbian soldiers cannot be discharged from the military if they openly declare their sexual orientation
   b. women cannot be excluded from participating in combat during state-declared military invasions
   c. male and female soldiers cannot be assigned to separate living quarters on military bases
   d. women cannot be denied admission to state-supported military academies

8. The organizational linkages created when a small number of people from elite groups sit on the Board of Directors of several companies, universities, and foundations are called _____.
   a. political action committees
   b. interlocking directorates
   c. marginalized groups
   d. national alliances

9. Which model of power interprets the state as its own major constituency that develops and perpetuates its own interests?
   a. pluralist
   b. power elite
   c. congressional
   d. autonomous state

10. One way that the state directs public opinion is through _____, or the restriction of people's access to certain information, such as sexually explicit materials on the Internet.
    a. coercion
    b. revolution
    c. censorship
    d. propaganda

11. According to Max Weber, which of the following is characteristic of bureaucracy?
    a. shared decision-making between managers and workers
    b. clear and specific division of labor
    c. vague policies and procedures
    d. close personal relationships

12. Under rational-legal authority, authority is gained through the _____.
    a. long-standing personal relationship the leader and his/her family has with the citizens of the state
    b. personal appeal of the leader, who is believed to have special gifts that inspire devotion
    c. process of being elected or appointed in accordance with a formal set of procedures
    d. use of physical force and other forms of coercion

13. Which of the following statements about the United States military is (are) **true**?
    a. The representation of women in the military has increased considerably since the elimination of the draft.
    b. Although Black men have a long history of military service, the armed services were segregated by race until 1973.
    c. The admission of women into military academies has led to a reduced emphasis on competition and hierarchy in the military.
    d. all of these choices

14. In sociological terms, the best predictors of whether someone in the United States will vote in an election are _____.
    a. age, gender, and marital status
    b. sex, occupation, and income
    c. age, income, and education
    d. race, ethnicity, and gender

15. Of these democratic nations, which country has the lowest rate of voter participation?
    a. United States
    b. South Africa
    c. Australia
    d. Japan

16. Which of the following economic systems is characterized by private ownership of the means of production, profit generated by the workers' production of the goods and services, and owners' disproportionate consumption of goods and profits?
    a. Communism
    b. Democracy
    c. Capitalism
    d. Socialism

17. American workers who feel anxious about unemployment in the rapidly changing global economy are particularly prone to _____, or the fear and hatred of foreigners.
    a. alienation
    b. xenophobia
    c. homophobia
    d. technophobia

18. The United States has experienced a transition from a predominantly goods-producing economy to an economy based on the provision of information and services, a process known as _____.
    a. deindustrialization
    b. underemployment
    c. downsizing
    d. inflation

19. Which of the following is a consequence of deskilling?
    a. Employees gain greater control over their tasks.
    b. Jobs become more complex and challenging.
    c. Employees are paid less for their work.
    d. all of these choices

20. Conflict theorists _____.
    a. view recent transformations in the workplace as the result of societal tension caused by power imbalances between social groups
    b. define work as a functional necessity for society because it integrates people and supports social order
    c. are primarily interested in how the workplace encourages the formation of social bonds
    d. explain wage inequality as an incentive system that makes people work harder

21. Which of the following racial-ethnic groups in the United States experiences the greatest unemployment today?
    a. Mexican Americans
    b. African Americans
    c. Native Americans
    d. Asian Americans

22. Occupational prestige ratings are based on the _____.
    a. S. Department of Commerce's ranking of the relative contribution of particular jobs to economic growth
    b. S. Bureau of the Census' ranking of the average wages paid for particular jobs
    c. S. Department of Labor's ranking of the objective value of particular jobs to society
    d. general American public's ranking of the value of particular jobs to society

23. Jane works in an auto parts store where her male coworkers hang Playboy magazine centerfolds in the lunchroom and tell jokes about the female body that make her very uncomfortable. Under federal law, this is an example of _____.
    a. *quid pro quo harassment*
    b. *hostile working environment harassment*
    c. *emotionally abusive harassment*
    d. none of these choices

24. It would cost a day's wages for an Indonesian woman to buy the Barbie doll that she helps make for distribution in the United States. This international division of labor is a component of the _____.
    a. military industrial complex
    b. global assembly line
    c. industrial revolution
    d. glass ceiling

25. Which of the following conditions is covered as a disability under the Americans with Disabilities Act of 1990?
    a. blindness
    b. pregnancy
    c. drug addiction
    d. all of these choices

True-False Questions

1. For people of all races, serving in the military leads to higher earnings relative to their non-military peers.
   TRUE or FALSE

2. The United States has one of the highest voter turnout rates among democratic nations.
   TRUE or FALSE

3. One example of the gender gap in politics is that men are more likely than women to vote as Democrat.
   TRUE or FALSE

4. Currently about one-third of military personnel are racial minorities.
   TRUE or FALSE

5. According to Weber, leaders that inspire devotion and obedience through personal appeal have traditional authority.
   TRUE or FALSE

6. Sexual harassment was first declared illegal by the 1986 Glass Ceiling Act, which identified harassment as a form of employment discrimination.
   TRUE or FALSE

7. A smaller percentage of American workers are located in the farming, fishing, and forestry category than in any other occupational category.

    TRUE or FALSE

8. Federal employment legislation does not cover same-sex harassment in the workplace, because by definition, sexual harassment is restricted to men's exploitation of women.

    TRUE or FALSE

9. Under the Americans with Disabilities Act (ADA), a reasonable accommodation for a student with a disability would be to excuse him from taking exams if he feels that test-taking is too emotionally or physically stressful.

    TRUE or FALSE

10. Both legal and illegal immigrants to the United States tend to have higher levels of education and occupational skills than the typical worker in their homelands.

    TRUE or FALSE

Fill-in-the-Blank Questions

1. The National Rifle Association (NRA), a constituency organized to promote its own agenda, is an example of a(n) _____ group.

2. One way that the state influences public opinion is by distributing _____ , or information disseminated by an organization that is intended to justify its own power.

3. The United States is a _____ , meaning that it is based on the principle of representing all people through the right to vote.

4. Those who do not hold regular jobs, but whose employment is dependent on demand, such as independent consultants, are _____ workers.

5. Jobs in the _____ labor market are characterized by relatively high wages, good benefits, greater stability, and opportunities for promotion.

Essay Questions

1. Describe and explain the patterns of political participation in the U.S. by race/ethnicity, social class, and gender.

2. Explain which theory of power you feel best explains the exercise of power in U.S. society and support your choice with evidence.

3. Describe the benefits and obstacles minorities, women, and gays/lesbians experience in the U.S. military

4. Discuss the patterns of employment, unemployment, and underemployment in the U.S. by race/ethnicity, gender and immigration status.

5. Discuss the impact of deindustrialization on U.S. workers.

Solutions

PRACTICE TEST

Multiple Choice Questions

1. D, 382 The police are responsible for local law enforcement. The military defends the nation against domestic and foreign conflicts. Prisons punish people who violate the law, which is created by the government.

2. A, 382 The government creates laws and procedures that help maintain order in society.

3. D, 384 Rational-legal authority, which is based on the election or appointment of leaders according to formal procedures, leads to the formation of bureaucracies.

4. B, 384 Monarchies derive their power from traditional authority, or long-established patterns of leadership.

5. A, 385-6 The pluralist model interprets power in society as coming from group representation of the diverse interests in society. The autonomous state model asserts that social order is maintained by an administrative system that supports the status quo. The power elite model views the state as representing the interests of a small ruling class [Table 14.1].

6. D, 387-8 Feminist theories of the state view state institutions as reflecting masculine values and protecting men's interests [Table 14.1].

7. D, 394 The Supreme Court case, *United States v. Virginia*, declared that state-supported military academies such as the Citadel can not exclude women.

8. B, 386 Interlocking directorates are created when people drawn from the same elite group serve on several boards of directors.

9. D, 387 The autonomous state model of power views the state as its own major constituency that works to maintain the status quo [Table 14.1].

10. C, 382 Censorship occurs when the state restricts the public's access to information.

11. B, 384 Bureaucracy is characterized by a hierarchy of authority, clear division of labor, explicit rules and procedures, and impersonality.

12. C, 384 Rational-legal authority is based on the election or appointment of leaders according to formal procedures. Traditional authority is derived from long-established patterns of leadership. Charismatic authority is derived from the personal appeal of a leader assumed to have special gifts.

13. A, 392-5 The United States armed forces were segregated by race until 1948. Women's participation in the military increased after the all-male draft was eliminated in 1973, but their participation has not altered the competitive nature or hierarchical structure of the institution.

14. C, 388-90 The social factors that best predict voting patterns in the United States are age, education, and income. People who are older, well educated, and in the middle or upper class are most likely to vote.

15. D, 389 Japan and the United States both have very low voter participation rates. Australia has a higher voter turnout for elections than all other democratic nations, and South Africa has the third highest turnout.

16. C, 396 Under capitalism, property and the means of production are privately owned, whereas industry is owned and managed by the state under communism. Democracy and monarchy are political, not economic, systems.

17. B, 396 Americans, particularly those in the working class who feel anxious about loss of employment due to economic restructuring, are prone to xenophobia, the fear and hatred of foreigners.

18. A, 397 Deindustrialization is the process whereby the United States has moved from a predominately goods-producing economy to a primarily post-industrial economy, which is based on the provision of information and services.

19. C, 398-9 The process of *deskilling* usually simplifies the work process, reduces workers' control over tasks, and decreases pay.

20. A, 400-1 Conflict theorists view transformations in the workplace as the result of inherent tensions in the social system caused by power differences between groups vying for economic and social resources. Symbolic interactionists focus on the formation of social bonds at work. Functionalists assert that work integrates people into society, and wage inequality acts as an incentive for people to work harder [Table 14.3].

21. C, 403 Native Americans have the highest unemployment in the United States due to inadequate training, inaccessible education, and considerably higher rates of disability than people from other racial-ethnic groups.

22. D, 405 Occupational prestige ratings are based on the general public's ranking of the value of particular jobs to the United States.

23. B, 407-8 Sexual harassment is the repeated, unwanted physical or verbal behavior of a sexual nature that occurs in the workplace. The law defines two forms of harassment: 1) *quid pro quo harassment* involves the exchange of sexual favors as a condition of employment, and 2) *hostile working environment* harassment occurs when the conditions of work are made uncomfortable for the worker due to verbal comments of a sexual nature and the display of sexualized objects.

24. B, 396 The global assembly line refers to the new international division of labor where research, development, and management are controlled by the most developed countries (such as the U.S., Japan, and Germany) and assembly line work is performed in the less developed countries (such as Indonesia), where materials and labor costs are low.

25. A, 409 The Americans with Disabilities Act (ADA), adopted by Congress in 1990, states that employers with fifteen or more employees are prohibited from discriminating against job applicants who are physically disabled or current employees who become physically disabled. A disability is defined as a condition or history of a condition that impairs a major life activity, such as work. Drug addiction and pregnancy are excluded from this federal law, but people who are blind or deaf are entitled to protection under this law.

True-False Questions

1. T, 393 People of all racial-ethnic groups who serve in the military have higher earnings relative to their non-military peers.

2. F, 389-90 As indicated in Figure 14.2, the U.S. has a very low rate of voter turnout compared to other industrialized countries.

3. F, 390 Women, Blacks, and Latinos are more likely than men and Whites to vote Democratic.

4. T, 393 Racial minorities are 34 percent of the military today: 20 percent African Americans, 8 percent Hispanic, and 6 percent other racial minorities.

5. F, 384 According to Weber, charismatic authority is derived from the strong personal appeal of a leader who is believed to have special gifts, which inspires obedience.

6. F, 408 Sexual harassment is considered a form of employment discrimination under the rules of the Civil Rights Act of 1964. The glass ceiling refers to the invisible barriers that prevent women and racial-ethnic minorities from being promoted to the highest levels of management in the workplace.

7. T, 406 Figure 14.4 indicates that there are fewer workers in the farming, fishing, and forestry occupational category than in all other categories.

8. F, 408 Sexual harassment is an abuse of power in the workplace regardless of the sex of the perpetrator and the victim. The law no longer distinguishes between same-sex and opposite sex parties in defining certain behaviors as harassment.

9. F, 409 Under the ADA, a reasonable accommodation could include altering the format of the test that the student must take, but it does not excuse students (or workers) from performing the essential tasks required of all students (or employees).

10. T, 406-7 Both legal and illegal immigrants tend to be better educated and more skilled than their poorest counterparts in their homeland.

Fill-in-the-Blank Questions

1. interest, 385
2. propaganda, 382
3. democracy, 388
4. contingent, 399
5. primary, 404

Essay Questions

1. Political Participation. The patterns of political participation in the U.S. are discussed on p. 388-390 in the text. Student responses should demonstrate understanding of the impact of race/ethnicity, social class, and gender.

2. Theories of Power. The theories of power are discussed on p. 385-388 in the text. Student responses should demonstrate understanding of political dynamics in the U.S.

3. Minorities in the U.S. Military. The benefits and obstacles minorities, women, and gays/lesbians experience in the U.S. military are discussed on p. 391-395 in the text. Student responses should demonstrate awareness of this experience and the reasons for it.

4. Employment Patterns. The patterns of employment, unemployment, and underemployment in the U.S. are discussed on p. 401-407 in the text. Student responses should demonstrate familiarity with these patterns, particularly as they apply in terms of race/ethnicity, gender, and immigration status.

5. Deindustrialization. Deindustrialization in the U.S. is discussed on p. 397-398 in the text. Student responses should demonstrate an understanding of the reasons for deindustrialization and how this process affects all levels of workers in the U.S.

POPULATION, URBANIZATION, AND THE ENVIRONMENT

BRIEF CHAPTER OUTLINE

CHAPTER FOCUS

This chapter examines demographic processes, theories, and patterns; urbanization and the lifestyles and residential patterns associated with it; and environmental issues, policies, and challenges.

QUESTIONS TO GUIDE YOUR READING

1. What are the major demographic issues related to the U.S. Census?

2. What are the three basic demographic processes, and how should they be understood in the context of diversity?

3. What are age-sex pyramids, and what demographic processes do they represent?

4. What is the significance of cohorts?

5. What are the strengths and weaknesses of the theories of population growth: Malthusian, demographic transition, and zero population growth?

6. What are the major issues involved with family planning and other population policies?

7. What are the major features of urban lifestyles, and what are the major racial and social class residential patterns in urban areas?

8. What are the major features of the human ecosystem and the major challenges confronting it?

9. How are environmental issues and policies affected by race and class?

10. What are the major concerns of ecological demography, and what is its utility in environmental policy?

KEY TERMS

(defined at page number shown and in glossary)

age-sex pyramid 418

cohort 419

crude death rate 416

demography 414

ecological globalization 435

environmental racism 431

human ecology 427

immigration 414

life expectancy 416

population density 426

KEY PEOPLE

(identified at page number shown)

CHAPTER OUTLINE

I. **DEMOGRAPHY AND THE U.S. CENSUS**

Demography studies the size, distribution, and composition of human populations. Demographers use the **census** and vital statistics to create a picture of the U.S. population. Two debates between the U.S. Census Bureau and the U.S. Congress concerning the year 2000 census were the use of the multiracial category and the use of probability sampling to reduce *undercounting* of the homeless, immigrants, minorities who live in ghetto neighborhoods, and others of low social status, who are most likely to be undercounted. Another body of data used in demography is **vital statistics**, which include information about births, marriages, deaths, and migrations in and out of the country.

II. **DIVERSITY AND THE THREE BASIC DEMOGRAPHIC PROCESSES**

The total number of people in society is determined by three variables: births, deaths, and migrations. These variables show different patterns by race, ethnicity, social class, and gender. **Immigration**, or migration into a society from outside, adds to the population, while **emigration**, the departure of people from a society, reduces the population. Population grows exponentially, with an upward accelerating curve.

A. <u>Birth Rate</u>
   1. The **crude birth rate** of a population is the number of babies born for every thousand members of the population each year.
   2. The crude birth rate reflects the *fertility* of a population, which is the number of live births per number of women in the population.

3. Fertility is different from *fecundity*, which is the potential number of children in a population that could be born per thousand women if every woman reproduced at her maximum biological capability during the childbearing years.

4. The birthrate for the entire world is about 27.1 per thousand people, while the overall birthrate for the U.S. is about 16.5 per thousand people. The rate varies according social factors such as racial-ethnic group membership, region, socioeconomic status, and religious affiliation.

   a. Racial-ethnic minority groups and lower-income groups tend to have somewhat higher birthrates than do White non-minority groups and middle- and upper-income groups.

   b. Religious and cultural differences are apparent in the birth rate; for example, Catholics have a higher birth rate than do non-Catholics of the same socioeconomic status.

B. Death Rate

1. The **crude death rate** of a population is the number of deaths each year per thousand people. Lower death rates are found in countries, and regions within nations, that have higher standards of living and better quality health care.

2. The **infant mortality rate**, or the number of deaths per year of infants less than one year old for every thousand live births, also reflects the standard of living in a country. For example, the overall infant mortality rate in the U.S. is about 11 deaths for every thousand live births, compared to about 35 deaths per thousand live births in developing countries such as Kenya. Factors that contribute to high infant mortality rates include: lack of adequate health care, little or no access to health care facilities, the presence of toxic wastes, and inadequate food supply.

3. The **life expectancy** of a population or group is defined as the average number of years that members of the group can expect to live. Life expectancy varies by gender, race-ethnicity, and social class. In the U.S., life expectancy has gone from 40 years of age in 1900 to 77.3 years of age today, which is lower than the average life expectancy in most other industrialized nations.

C. Migration

Migration can occur between nations as well as within the boundaries of a country. In the 1980s, for example, extensive internal migration by African Americans, Hispanics, Asian Americans, and Pacific Islanders occurred within the U.S. Migration patterns are often linked to employment availability in particular industries, resulting in an increase in people of color in urban and suburban areas.

III. **POPULATION CHARACTERISTICS**

The composition of a society's population can reveal a tremendous amount about the society's past, present, and future. The demographic data of a society are a record of its national history. The important data that sociologists investigate include a population's sex ratio, age composition, age-sex population pyramid, and age cohorts.

A. Sex Ratio and the Population Pyramid

1. The **sex ratio** is the number of males per one hundred females. A sex ratio above 100 means that there are more males than females in the population, while a ratio below 100 indicates that there are more females than males in the population.

2. In almost all societies, there are more boys born than girls, but because males have a higher infant mortality rate and a higher death rate after infancy, there are usually more females than males in the overall population.

3. The *age composition* of the U.S. population is currently undergoing major changes as more people are entering the 65-and-over age bracket. As the population ages, its older members will have more influence on national policy and a greater impact on health care, housing, and other areas where the elderly have traditionally experienced age discrimination.

4. Sex and age data can be combined in a graphical format called an **age-sex pyramid**, which reflects the birth rate of a society [See Figure 15.2].

B. Cohorts

A birth cohort, or more simply, a **cohort**, consists of all the people born within a given period. The baby boom cohort, born between 1946 and 1964, now comprises about one third of the entire United States population. This cohort has had a major impact on the practices, policies, habits, preferences, and culture of American society.

IV. THEORIES OF POPULATION GROWTH LOCALLY AND GLOBALLY

There is no scientific consensus on the planet's *carrying capacity*, or the number of people the planet can support on a sustained basis.

A. Malthusian Theory
1. Like other animals, humans can survive and reproduce only when they have access to the means of *subsistence*, or necessities of life, such as food and water.

2. **Malthusian** theory represents **Thomas Malthus'** idea that populations tend to grow faster than the subsistence needed to sustain them. Rather than adding the same number of individuals each year, populations tend to grow by *exponential increase*, in which the number of individuals added each year grows.

3. Malthus asserted that there were three major *positive checks* on population growth: famine, disease, and war. These checks inevitably come into play when populations rise to the level of subsistence and slightly beyond.

4. Malthusian theory actually predicted rather well the population fluctuations of many agrarian societies such as Egypt from about 500 AD through Malthus' own lifetime. However, Malthus failed to foresee three revolutionary developments that derailed his predicted cycle of growth and catastrophe.
   a. In agriculture, technological advances have permitted farmers to work larger plots of land and grow more food per acre, resulting in higher subsistence levels than Malthus predicted.
   b. In medicine, scientific developments have treated diseases that Malthus expected to wipe out entire nations, such as the bubonic plague.
   c. The development and widespread use of contraceptives in many places has kept the birth rate lower than Malthus would have thought possible.

B. Demographic Transition Theory

An alternative to Malthusian theory is **demographic transition theory**, which proposes that countries pass through a consistent sequence of population patterns linked to the degree of development in the society [See Figure 15.3]. The three main stages of population change according to this theory are:

1. Stage 1: characterized by a high birth rate and high death rate.
2. Stage 2: characterized by a high birth rate but a low death rate, resulting in an overall population increase.
3. Stage 3: characterized by a low birth rate and low death rate, resulting in stabilization of the overall level of the population as medical advances continue and general cultural changes occur, such as a smaller ideal family size.
4. Demographic transition theory has been criticized as being *ethnocentric* (based on models from highly industrialized, predominantly White Societies).

C. The "Population Bomb" and Zero Population Growth
   1. In 1968, Paul Ehrlich noted that worldwide population growth has outgrown food production, and that massive starvation must inevitably follow.
   2. Ehrlich was among the earliest thinkers in modern times to argue that the quality of the environment, especially the availability of clean air and water, was a critical factor in the growth and health of populations.
   3. Many of the dire predictions Ehrlich made have come true, including mass starvation in parts of Africa and starvation among some Black and Hispanic populations in the U.S.; increased homelessness in American cities, especially among minorities; acid rain; and extinction of plant and animal species.
   4. Organizations such as Zero Population Growth (ZPG) are dedicated to reaching the **population replacement level**, a condition in which the combined birth and death rate of a population sustains the population at a steady level. By 1980, the U.S. had reached this level of reproduction, partly due to the use of birth control.

V. **CHECKING POPULATION GROWTH**

By the 1980s, countries representing 95 percent of the world's population had formulated policies aimed at stemming population growth, although there is no consensus on how population growth should be controlled. Efforts to encourage the use of contraceptives, for example, have had mixed support, because some political and religious groups are opposed to the use of birth control.

A. Family Planning and Diversity
   1. Many governments make contraceptives available to individuals, but this is not always consistent with the beliefs and cultural practices of all groups in the society. Governmental programs that advocate contraception can only be successful if couples themselves choose smaller families over larger ones.
   2. Birth rate and family size are correlated with the overall level of economic development of a country, as well as the economic status of certain ethnic groups within a country, as reflected in the lower birth rates and smaller average family size in more economically developed nations.
   3. The assumed relationship between economic development and family size in demographic transition theory has been challenged. In Bangladesh, for example, the population has

become receptive to birth control programs, thereby lowering the birth rate in the absence of Western-style economic development.

B. Population Policy and Diversity

Family planning programs offer great potential for achieving significant declines in birth rates. In overpopulated countries where such programs can have the most effect, the demand for family planning resources surpasses the supply. In the U.S., there is some cultural resistance on the part of some racial and ethnic groups to government-sponsored contraceptive programs due to fears that widespread use of contraception threatens the very survival of these groups. Such governmental programs may be perceived as racist.

VI. URBANIZATION

The growth and development of cities, or centers of human activity with a high degree of population density, is a relatively recent occurrence in the course of human history. The extent to which a community has the characteristics of city life is referred to as **urbanization**.

A. Urbanization as a Lifestyle
 1. **Georg Simmel** (1950/1905) argued urban living has profound social psychological effects on the individual, including insensitivity to the people and events around him or her due to the intensity of city life, and discouragement of close, personal interaction. **Emile Durkheim** had also noted that the suicide rate was greater in more urbanized areas than in rural areas.
 2. **Louis Wirth** (1938) agreed that the city was a center of distant, cold impersonal interaction, resulting in the urban dweller experiencing alienation, loneliness, and powerlessness. However, both Simmel and Wirth believed that city life, with its relative absence of close, restrictive ties, offered individuals a feeling of freedom.
 3. Herbert Gans provided a contrasting view of urban life, concluding that many city residents have strong loyalties to others and develop a sense of community. Gans classified the *urban village* as having several "modes of adaptation."
    a. *Cosmopolites* are typically students, artists, writers, and musicians who choose urban living to be near the city's cultural facilities, and together, form a tightly knit community.
    b. *Ethnic villagers* live in ethnically and racially segregated neighborhoods.
    c. The *trapped* are individuals who, similar to today's *urban underclass*, are unable to escape from the city because of extreme poverty, homelessness, unemployment, and other familiar urban problems.

B. Race, Class, and the Suburbs
 1. The impact of race and class is clear in the distinction between city and suburb, as only about one-fourth of African Americans live in suburban areas today.
 2. In the suburbs, one chooses one's neighbors and friends on the basis of race and educational and occupational similarity. Segregation in interpersonal interaction is encouraged by practices that support residential segregation.
 3. People of color, particularly African Americans, often become as segregated in suburbs as in cities due to the practices of White landlords, homeowners, and realtors, who may selectively show particular properties only to people of a certain racial group.

4. Banks that practice *redlining* make it virtually impossible for persons of color to get a mortgage loan for a specific property, further intensifying residential segregation.

C. The New Suburbanites

1. One consequence of the 1924 National Origins Quota Law was that it encouraged immigration from Northern and Western Europe, while discouraging additional immigration from Eastern and Southern Europe.

2. Today, the most prominent immigrants in suburban neighborhoods are Hispanics and Asian Americans. Suburban Whites may perceive themselves to be in competition for jobs and housing with these new immigrant groups.

3. Current immigration has led to what William Frey called the new *demographic divide*, which refers to settlement that is concentrated in small cities and suburbs on the East and West coasts, with little effect on the other parts of the country.

VII. ECOLOGY AND THE ENVIRONMENT

It should be apparent that population size has an important social dimension. Social forces can cause changes in the size of a population, and population changes can transform society. **Population density** is the number of people per unit of area, usually per square mile. As population density rises to high levels, the familiar problems of urban living appear, including high rates of crime and homelessness. Interacting with these problems are crises of the physical environment, such as air and water pollution, acid rain, and a growth in hazardous waste output.

Human ecology is the scientific study of the interdependencies between humans and their physical environment. A **human ecosystem** is any system of interdependent parts that involves human beings interacting with each other and the physical environment. Two fundamental, closely related problems confront our present ecosystems—overpopulation and exhaustion of natural resources.

A. Vanishing Resources

In all ecosystems, organisms depend upon each other and the physical environment for survival. The supply of many natural resources is finite. If one element in an ecosystem is disturbed, the entire system is affected, as exemplified by the problems that emerged in the 1940s and 1950s with the use of DDT in the U.S. Whereas growing population is a problem of the developing world, shrinking resources are problematic in the industrialized world, where real estate development takes over millions of acres of farmland each year in the United States alone.

B. Environmental Pollution

1. The most threatening forms of pollution are the poisoning of the planet's air and water, with the U.S., Japan, and Russia acting as leading air and water polluters.

2. On the industrial side, the Environmental Protection Agency (EPA) estimates that the hazardous and cancer-causing pollutants released into the air by industry are responsible for about 2,000 deaths a year. When mixed with other chemicals normally present in the air, sulfur dioxide turns into sulfuric acid, which gets carried back to earth in droplets of *acid rain*.

3. A daunting international issue has grown around a group of chemicals called chlorofluorocarbons (CFCs), which are used in the manufacture of plastics, as a coolant in refrigerators, and as an aerosol propellant.

4. Related to the problem of ozone depletion is the **greenhouse effect**, which results in small changes in the average temperature of the earth that can have dramatic consequences. For example, it can cause melting in the arctic regions, which raises the level of the sea, affecting water, land, and weather systems worldwide.

5. Only a very small portion of the earth's water is usable by humans. The nation's rivers and lakes have been dumping grounds for heavy industry, yet these same industries (paper, steel, automobile and chemical) depend upon clean water for their production processes, during which they take water from the rivers and lakes and return it heated and polluted. The difference in temperature can alter aquatic habitats and kill aquatic life, earning it the name **thermal pollution**.

6. The EPA estimates that 63 percent of rural Americans may be drinking water that is contaminated as a result of agricultural runoff and improper disposal of toxic substances in landfills. Polluting continues despite legislative prohibitions.

C. Environmental Racism and Classism

Toxic wastes are stored and dumped with disproportionate frequency in areas that have high concentrations of racial-ethnic minorities, a phenomenon known as **environmental racism**. [See Map 15.2] For example, the largest commercial hazardous waste landfill in the U.S. is in Emelle, Alabama, an area where 80 percent of the population is Black.

D. Feminism and the Environment

Women generally feel more vulnerable than men to risks posed by environmental problems. Consequently, they show greater concern with issues of environmental risk. Lack of attention on the part of local and federal governments to environmental issues can be interpreted as lack of attention to policy that differentially affects women.

E. Environmental Policy

U.S. environmental policy has been affected by an organized environmental movement, which brought increasing attention to environmental hazards and threats. In the past three decades, federal and local agencies have better managed the problems of environmental pollution through stiffer antipollution laws and the encouragement of alternative technologies. However, industry has resisted antipollution laws because of the costs, and unions have resisted such laws for fear of job loss.

VIII. GLOBALIZATION: POPULATION AND ENVIRONMENT IN THE TWENTY-FIRST CENTURY

Sociologists predict that the U.S. will experience increasing suburbanization, with accompanying increases in heavy industry and thus additional pollution. Sociologists are concerned with the mutual effects of lifestyle and a changing planet. Ecological concerns stimulated the development of **ecological demography**, a field combining demography and ecology to explore alternative technologies. **Ecological globalization** has resulted in increasing attention to environmental problems on the international political agenda, such as concern about *genetically modified organisms* (GMOs).

INTERNET EXERCISES

1. The website of the Environmental Protection Agency (www.epa.gov/) has extensive databases, information on its programs, laws and regulations, and publications. For a fun and informative exercise, use the interactive mapping exercise entitled "Window to My Environment," through which you can identify resources like watersheds, and problems, like superfund sites. Complete the exercise, print out a map and write a short analysis, indicating what surprised you and how safe you now feel in your neighborhood.

2. World Resources Institute (www.wri.org/) publishes a range of reports and multimedia presentations on global environmental and population issues. It is also an advocacy group, presenting ways in which people can take action. Select an issue that interests you from the major categories like agriculture and food, biodiversity, climate change, population and health, or water resources. Report on your findings. This would be a great resource if you are doing a term project on an issue in this area.

3. Population Reference Bureau (www.prb.org/) has extensive information on population topics in the US, including a state-by-state database, PowerPoint presentations, fact sheets and more. It also publishes reports on world population trends and issues. Depending on your course assignments, write a short report on a topic that interests you, or use the site for a longer project on the same topic.

INFOTRAC EXERCISES

Conduct a keyword search using InfoTrac College Edition to extend the discussion and exercise, TAKING ON SOCIAL ISSUES: *Environmental Racism* (p. 431).

1. *Keyword: Environmental racism.* This search leads to articles linking environmental racism to genetic engineering, labor, climate change, sewage plans, toxic dumping, the environmental and global justice movements, and legal cases.

2. *Keyword: Environmental justice.* Searching with this keyword yields a large number of articles on poverty and pollution, chemical industry accidents, garbage wars in cities, civil/human rights and the environment, the EPA, public policy and lawsuits.

3. *Keyword: Environmental equity.* This search leads to a number of articles on sustainable planning, social costing, distributing environmental quality, regionalism, hazardous and toxic waste disposal, neighborhoods, and environmental movements.

PRACTICE TEST

Multiple Choice Questions

1. Which of the following statements about the census is (are) **true**?
   a. The U.S. constitution requires that the census be conducted every ten years.
   b. The U.S. census uses probability sampling to estimate the population because counting every individual is too expensive and time-consuming.
   c. The U.S. census collects information about people's opinions on key social issues.
   d. all of these choices

2. The scientific study of the size, composition, and distribution of the population is _____.
   a. epidemiology
   b. demography
   c. urbanism
   d. ecology

3. The measure of the potential number of children in a population that could be born per thousand women, if every woman reproduced at her maximum biological capacity during the childbearing years, is the _____ rate.
   a. fertility
   b. fecundity
   c. mortality
   d. maternalism

4. The major restructuring of the age-sex pyramid that will occur in the U.S. over the next twenty-five years is due to the _____.
   a. considerable increase in infant mortality among minority populations
   b. significant increase in the immigration of Hispanics to the country
   c. recent explosion in the birth rate for Whites
   d. aging of the Baby Boom cohort

5. Who was one of the first modern theorists to argue that the quality of the environment, such as the availability of clean air and water, is a critical factor in the health of the population?
   a. Paul Ehrlich
   b. Herbert Gans
   c. Kingsley Davis
   d. Thomas Malthus

6. Which of the following industrialized nations has the lowest infant mortality rate?
   a. China
   b. Japan
   c. Russia
   d. United States

7. Which of the following factors did Malthus surmise was a positive check on population growth?
   a. immunizations
   b. abortion
   c. disease
   d. all of these choices

8. The state in which the combined birth and death rate of a population simply sustains the population at a steady level is the _____ level.
   a. positive check
   b. limited growth
   c. population explosion
   d. population replacement

9. Which of the following statements about life expectancy in the U.S. is (are) **true**?
   a. Average life expectancy in the U.S. increased from 40 years old to 77 years old in the period between 1900 and 2000.
   b. Average life expectancy in the U.S. is higher for men than for women and higher for White people than for Black people.
   c. Average life expectancy is higher in the U.S. than in all other industrialized nations.
   d. all of these choices

10. The age-sex pyramid of Mexico in 1990 visually resembles which shape?
    a. horizontal rectangle
    b. vertical rectangle
    c. triangle
    d. square

11. Simmel and Wirth have both argued that urban life has what social-psychological effect(s) on individuals?
    a. feeling of liberation or freedom from restrictions
    b. heightened sensitivity to events surrounding them
    c. increased opportunity for close, personal interaction
    d. all of these choices

12. In Gans' vision of the urban village, which of the following groups choose urban living to be near the city's cultural facilities?
    a. urban underclass
    b. ethnic villagers
    c. cosmopolites
    d. the trapped

13. The examination of ecosystems has demonstrated that _____.
    a. the supply of natural resources is finite
    b. humans and the physical environment are interdependent
    c. a disturbance in one element of an ecosystem has an effect on the entire system
    d. all of these choices

14. Studies of exposure to environmental toxins have found which of the following?
    a. American Indians, Hispanics, and African Americans as well as people of lower socioeconomic status reside disproportionately closer to toxic sources than do Whites and people of higher socioeconomic status.
    b. Environmental toxins are pretty evenly distributed around the country, with people of all races and income levels equally exposed.
    c. Socioeconomic status, not race or ethnicity, is the greatest predictor of exposure to environmental toxins.
    d. Environmental toxins are a problem only in large urban areas, while rural areas in the U.S. are virtually toxin-free.

15. Which demographic theory is fundamentally pessimistic, predicting that the population will ultimately outstrip the food supply, resulting in worldwide starvation and rampant disease?
    a. Malthusian theory
    b. Doomsday theory
    c. Rational Choice theory
    d. Demographic Transition theory

16. Stage two of the demographic transition is characterized by a _____.
    a. low birth rate and low death rate
    b. declining birth rate and high death rate
    c. high birth rate and declining death rate
    d. high birth rate and high death rate

17. The internal migration patterns of Hispanics in the U.S. have been strongly linked to what factor?
    a. desire to live in sparsely populated, rural areas
    b. availability of bilingual educational facilities
    c. presence of Catholic Churches
    d. availability of employment

18. To reach zero population growth, the average number of children per family in a society would have to be _____.
    a. one
    b. two
    c. three
    d. zero

19. Vital statistics include which of the following types of information?
    a. births and deaths
    b. marriages and divorces
    c. immigration and emigration
    d. all of these choices

20. In the United States, babies of which racial-ethnic group are *most* likely to contract Acquired Immune Deficiency Syndrome (AIDS)?
    a. African American
    b. Chinese American
    c. Mexican American
    d. European American

True-False Questions

1. In almost all societies, there are more girls than boys born, resulting in a greater number of females than males in all age groups.

   TRUE or FALSE

2. Israel has experienced considerable population growth since 1948 due to the migration of Jews from Europe and the United States.

   TRUE or FALSE

3. The United States was in the first stage of the demographic transition during the industrialization period of the late 1800s.

   TRUE or FALSE

4. The estimated undercount of African Americans in the 2000 U.S. Census is about 2 percent.

   TRUE or FALSE

5. The practice of *redlining* by banks contributes to racial integration in the suburbs by allocating a certain percentage of mortgages to African Americans who want to purchase housing in predominantly White neighborhoods.

   TRUE or FALSE

6. Because stricter environmental laws were passed in the last two decades, the United States is no longer a major contributor to air and water pollution around the world today.

   TRUE or FALSE

7. According to Herbert Gans, *cosmopolites* are city residents who are trapped in urban areas due to extreme poverty and unemployment.

   TRUE or FALSE

Fill-in-the-Blank Questions

1. A human _____ consists of the interdependent forces of human population, natural resources, and the condition of the physical environment.

2. The departure of people from a society, or _____ , results in a decrease in population.

3. The group of approximately 75 million babies born in the U.S. between 1946 and 1964, which now represents nearly one-third of the population, is known as the _____ .

4. Population _____ , which usually refers to the number of people per square mile, is higher in urban areas than in rural areas.

5. The crude birth rate reflects the _____ of a population, which is the number of live births per number of women in that population.

Essay Questions

1. Discuss the demographic transition theory and recent challenges to it.

2. Discuss how and why life expectancy in the U.S. varies by sex, race, and social class.

3. Explain how racism and classism intersect with environmental pollution in the United States, citing a specific example of environmental racism that has been documented by researchers.

4. Explain why there might be cultural resistance to government-sponsored family planning programs among certain racial-ethnic minority groups in the United States.

5. Who are the "new suburbanites" and what issues does their immigration present for social institutions in the United States?

Solutions

PRACTICE TEST

Multiple Choice Questions

1. A, 414 The U.S. census, conducted every ten years, attempts to count every individual and collect basic demographic information such as race, marital status, and income. The Census does not ask people for their opinions on social issues.

2. B, 415 Demography is the scientific study of the population. Epidemiology is the study of all factors associated with disease. Ecology is the study of the physical environment. Urbanism is the extent to which a community has the characteristics of city life.

3. B, 415 Fecundity refers to the potential number of children that could be born if every woman reproduced at her maximum capacity, while fertility is the number of live births per number of women in the population. The crude death rate is the number of deaths each year per thousand people, while the infant mortality rate refers to the number of deaths per year of infants under one year old for every thousand births.

4. D, 418 The aging of the Baby Boom cohort, a trend called the "graying of America," has resulted in major restructuring of the age-sex pyramid in the United States.

5. A, 423 In the 1960s, Paul Ehrlich noted that the quality of the environment is related to the health of the population. Ehrlich advocates reducing population growth until the population replacement level is attained.

6. B, 416 Table 15.1 shows that Japan has the lowest infant mortality rate (at 3.9), while the U.S. has a rate of 6.8, China 28.1, and Russia have the highest infant mortality rates of all industrialized nations.

7. C, 420-1 Malthus argued that war, disease, and famine were positive checks on population growth. Immunizations increase population by lowering the death rate. Writing in the 1700s, he could not foresee the contributions of abortion and artificial birth control to reducing population growth.

8. D, 423 The state in which the combined birth and death rate of a population sustains the population at a steady level is the population replacement level.

9. A, 416 Average life expectancy in the United States has increased to 77 years old, which is considerably longer than in 1900, but lower than the average life expectancy in nearly every other industrialized nation. Within the U.S., life expectancy is lower for men, members of racial-ethnic minority groups, and lower-income people.

10. C, 419 Figure 15.2 depicts the age-sex pyramid of Mexico as a triangular shape. The pyramid of the U.S. resembles a vertical rectangle with a bulge in the middle.

11. A, 425 Both Simmel and Wirth recognized that city life could be personally liberating, but it is also characterized by cold, impersonal relationships, decreased sensitivity to one's surroundings, and feelings of alienation and powerlessness.

12. C, 425 In Gans' vision of the urban village, cosmopolites consist of students, artists, and musicians who live in the city to be near cultural facilities. Ethnic villagers live in segregated neighborhoods, or ethnic enclaves. The trapped are similar to today's urban underclass, consisting of those people who are unable to escape from the city due to poverty.

13. D, 427 The examination of ecosystems reveals that the supply of many natural resources is finite, humans and the physical environment are interdependent, and a disturbance in one part of the ecosystem has an effect on the whole system.

14. A, 432-3 American Indians, Hispanics, and African Americans as well as people of lower socioeconomic status reside disproportionately closer to toxic sources than do Whites and people of higher socioeconomic status. Race and ethnicity are the strongest predictors of exposure to environmental toxins, which occur in both urban and rural areas.

15. A, 420-1 Malthusian theory suggests that the population will eventually surpass the food supply despite positive checks on growth.

16. C, 421-2 Stage two of the demographic transition is characterized by a high birth rate and declining death rate. Stage one is characterizes by high birth and death rates, while low birth and death rates are characteristic of stage three.

17. D, 417 Internal migration patterns of Hispanics have traditionally been linked to the agricultural industry, and more recently, to the meatpacking and textile industries located near urban centers; thus, migration is strongly influenced by the availability of employment.

18. B, 423 To reach population replacement level, which would result in zero population growth, each family has to have an average of two children.

19. D, 414 The vital statistics for a nation include the number of births, deaths, marriages, divorces, and migration into (immigration) and out of (emigration) the country.

20. A, 417 African American babies are almost thirty times more likely than White babies to contract AIDS, and Hispanic babies are twenty-five times more likely to contract the disease, contributing to lower infant survival rates for these groups compared to White European Americans.

True-False Questions

1. F, 418 In almost all societies, more boys than girls are born, but boys and men have higher mortality rates, resulting in more women than men in adulthood.

2. T, 417 Israel has experienced population growth due to the migration of Jews from Europe and the United States. Those who migrate are more likely to be younger adults.

3. F, 421-2 The United States was in stage one of the demographic transition during the Colonial period and stage two during Industrialization. It is currently in stage three.

4. F, 414 While the estimated overall undercount on the 2000 U.S. Census was 2 percent, the estimated undercount of African Americans was as high as 20 percent.

5. F, 426 The practice of *redlining* is a form of discrimination whereby banks deny

6. F, 429 Despite stricter legislative prohibitions on pollution, the United States continues to be one of the largest contributors in the world to air and water pollution.

7. F, 425 According to Gans, cosmopolites tend to be students and artists who choose urban

living to be close to cultural facilities. Ethnic villagers live in segregated neighborhoods, or ethnic enclaves. The trapped are similar to today's urban underclass, consisting of those people who are unable to escape from the city because of unemployment and extreme poverty.

Fill-in-the-Blank Questions

1. ecosystem, 427
2. emigration, 415
3. Baby Boomers, 419
4. density, 426
5. fertility, 415

Essay Questions

1. Demographic Transition Theory. The demographic transition theory is discussed on p. 421-422, and challenges to it on p. 424-425. Student responses should demonstrate an understanding of the theory and the basis for the challenges to it in developing countries.

2. Life Expectancy. Life expectancy on the basis of sex, race, and social class in the U.S. is discussed in the text on p. 415-417. Student responses should demonstrate and understanding of how life expectancy is calculated and the reason for differences on the basis of sex, race, and social class.

3. Environmental Racism. The intersection of racism and classism with environmental pollution in the U.S. is discussed on p. 431-433. Student responses should demonstrate an understanding of the patterns of environmental pollution in the U.S. and how these are related to race and class.

4. Family Planning. Cultural resistance to government-sponsored family planning among certain racial-ethnic minority groups in the United States is discussed on p. 423-424. Student responses should demonstrate an understanding of the distrust minorities have of government programs in generation, and their specific concerns about family planning.

5. The New Suburbanites. The "new suburbanites" are discussed on p. 426 in the text. Student responses should demonstrate an understanding of the immigration patterns that have led to the new suburbanites, and how these patterns present challenges to U.S. social institutions.

SOCIAL CHANGE AND SOCIAL MOVEMENTS

BRIEF CHAPTER OUTLINE

CHAPTER FOCUS

This chapter examines the types, causes, theories, and consequences of social change; the forms of collective behavior; and the types, causes and theories of social movements. The effects of globalization and diversity on social change are also explored.

QUESTIONS TO GUIDE YOUR READING

1. What is social change and what are its major characteristics?

2. What are the major theories of social change?

3. What are the causes of social change?

4. What are the social consequences of modernization?

5. What are the major global theories of social change, and what are their strengths and weaknesses?

6. What is collective behavior and what are its characteristics?

7. What are the main types of social movements?

8. Under what conditions to social movements emerge?

9. What are the major theories explaining the development of social movements, and what are their strengths and weaknesses?

10. What are the effects of globalization and diversity on social change?

KEY TERMS

(defined at page number shown and in glossary)

collective behavior 453-4

cyclical theories 443

evolutionary theory 441

gemeinschaft 449

globalization 460

macrochange 440

mobilization 457

modernization theory 452

multidimensional social change evolutionary theory 442

other-directedness 452

political process theory 461

KEY PEOPLE

(identified at page number shown)

CHAPTER OUTLINE

I. WHAT IS SOCIAL CHANGE?

Social change—the alteration of social interactions, institutions, stratification systems, and elements of culture over time—includes both **microchanges** (like fads) and **macrochanges** (like *modernization*). All social changes share the following characteristics: **Social change is uneven**, and may involve *cultural lag*, a term coined by **William F. Ogburn** (1922). **The onset and consequences of social change are often unforeseen. Social change often creates conflict**, in terms of race-ethnicity, social class, and gender. **The direction of social change is not random**, and may be wanted or resisted.

II. THEORIES OF SOCIAL CHANGE

A. Functionalist and Evolutionary Theories

1. Functionalist theorists, like **Herbert Spencer** and **Emile Durkheim**, build upon the belief that all societies, past and present, possess basic elements and institutions that perform certain *functions* that permit a society to survive and persist. They assume that as societies move through history, they become more complex.

2. Emile Durkheim identified two types of societies:
    a. Foraging and pastoral societies are structurally simple, homogeneous societies based on *mechanical solidarity*, where members engage in largely similar tasks.
    b. Agricultural, industrial, and post-industrial societies are structurally more complex, heterogeneous societies based on *organic* or *contractual solidarity*, where great social differentiation exists and there is extensive division of labor among people who perform many specialized tasks.

3. **Evolutionary theories** of social change are a branch of functionalist theory.
    a. **Unidimensional evolutionary theory**, now out of favor, argued that societies follow a single evolutionary path from "simple," relatively undifferentiated societies to more complex societies, which were perceived as more "civilized."
    b. **Multidimensional evolutionary theory** argues that the structural, institutional, and cultural development of a society can follow many evolutionary paths simultaneously, with the different paths emerging from the circumstances of the society.
        i. The nature of social evolution in the society depends upon the interplay between the society's technology, population characteristics, extent of social differentiation, and other structural and cultural elements.
        ii. Gerhard Lenski and colleagues posit that technology has a central role in development because technological advances are significantly, although not wholly, responsible for other changes, including the nature of law, alterations in religious preference, form of government, and relations between diverse groups.

4. The theory sees change as evolutionary, cumulative, and not easily reversible, but it does acknowledge instances of social reversal. Newer functionalist theories emphasize the role of racial, ethnic, social class, and gender differences in the process of social change.

B. Conflict Theories
    1. The central notion of conflict theory is that conflict is built into social relations. For **Karl Marx**, social conflict, particularly between the two major social classes in any society (proletariat and bourgeoisie) was the driving force behind all social change.
    2. Sociologists now think that conflict between Whites and racial-ethnic minorities is at least partly rooted in class conflict, because there are important cultural differences *within* broadly defined racial-ethnic categories, and minorities are disproportionately represented among the lower-income classes.
    3. A central feature of Marx's work is that revolution and dramatic social change will come about when class conflict inevitably leads to a decisive social rupture. Although the worldwide revolution predicted by Marx never developed, his analysis of class-related conflict has advanced understandings of social change.
    4. Marx seems to have overemphasized the role of economics and ignored the importance of other relevant social factors in the social tensions he observed. For example, Theda Skocpol has

noted that in France, Russia, and China, countries where major revolutions have occurred, serious internal conflicts between classes were combined with major international crises that the elite classes proved unable to resolve before they were overthrown.

5. Ethnic, racial, and religious differences join social class differences as major causes of conflict within and between countries. Examples of this include recent bouts of so-called *ethnic cleansing*, in which one ethnic group attempts to annihilate another, as in Bosnia in the 1990's.

C. Cyclical Theories

1. **Cyclical theories** of social change invoke patterns of social structure and culture that are believed to recur at more or less regular intervals.

2. Arnold Toynbee, social historian and a principal theorist of cyclical social change, argues that societies are born, mature, decay, and sometimes die.

3. Sociological theorists **Pitrim Sorokin** and Theodore Caplow have argued that societies proceed through three different phases or cycles.

    a. In the first phase, called *idealistic culture*, the society wrestles with the tension between the ideal and the practical.

    b. The second phase, *ideational culture*, emphasizes faith and new forms of spirituality.

    c. The third phase is *sensate culture*, which stresses practical approaches to reality and involves the hedonistic and sensual elements of culture.

III. THE CAUSES OF SOCIAL CHANGE

A. Revolution

A **revolution** is the overthrow of a state or the total transformation of central state institutions. Revolutions can sometimes break down a state as a result of conflict between an oppressive state and various disenfranchised groups, who are dissatisfied and have the opportunity to mobilize en mass. Structured opportunities may be created through war or an economic crisis, or mobilization through a *social movement*. Social structural conditions that often lead to revolution include a highly repressive state, a major economic crisis, or the development of a new economic system.

B. Cultural Diffusion

Cultural diffusion is the transmission of cultural elements from one society or cultural group to another. Diffusion of cultural elements, as **Ralph Linton** demonstrated, can occur by means of trade, migration, mass communications media, and social interaction. For example, many contemporary practices among Black Americans, such as step shows, are traceable to Africa.

C. Inequality and Change

Inequalities between groups of people on the basis of sex, class, race, ethnicity, or other social structural characteristics can be a powerful spur toward social change. Culture itself can sometimes contribute to the persistence of social inequality, therefore, it may become a source of discontent for certain groups in society.

D. Technological Innovation and Cyberspace Revolution

Technological innovations and inventions can be strong catalysts of social change. The path by which technology is introduced into society, and the impact it has on the lives of various groups, often reflects the predominant cultural values in that society. The cyberspace revolution has moved most of the globe from a *culture of calculation* to a *culture of simulation*.

E. Population and Change
   1. Limitations placed on the population by the natural environment can greatly influence the nature of social relationships. For example, crowding affects how people interact with each other.
   2. Major social change can also result from shifts in the age composition of a population. For example, as the average age in the United States increases, the society must respond to the needs and desires of the aging population.
   3. Immigration is having profound effects on the ethnic and racial composition of the United States. By the year 2050, it is expected that Hispanics will comprise 25 percent and Asians will comprise 9 percent of the U.S. population, which will influence the structure of the economy, the ethnic mix in education and employment, and styles of dance, music, and language.

F. War, Terrorism, and Social Change

   War, severe political conflict, and terrorism result in large and far-reaching changes for both conquering and conquered groups.

IV. MODERNIZATION

Modernization is a process of social and cultural change that is initiated by industrialization and followed by increased social differentiation and division of labor. The process of modernization is characterized by the *decline of small, traditional communities as a society becomes more bureaucratized* and interactions come to be shaped by formal organizations. There is generally *a decline in the importance of religious institutions*, and with the mechanization of daily life, people often feel that they have lost control of their lives.

A. From Community to Society

   The German sociologist **Ferdinand Tonnies** viewed the process of modernization as a progressive loss of **gemeinschaft**, which is German for "community." A gemeinschaft is characterized by a sense of fellowship, strong personal ties and primary group memberships, and a sense of personal loyalty to one another.

   1. Tonnies argued that urbanization and the industrial revolution, with its emphasis on efficiency and task-oriented behavior, destroyed the sense of community and personal ties associated with an earlier rural life, substituting instead feelings of rootlessness and impersonality.
   2. These changes resulted in the condition of **gesellschaft**, which is German for "society." This kind of social organization is characterized by a high division of labor, less prominence of personal ties, the lack of a sense of community among the members of society, and the absence of a feeling of belonging.

B. Mass Society and Bureaucracy
   1. Modernization has produced what is called a *mass society*, or one in which industrialization and bureaucracy reach exceedingly high levels and the change from gemeinschaft to gesellschaft is accelerated.

2. The breakup of primary ties is particularly pronounced in a mass society, where the government expands to include tasks that were previously done by the family, and it becomes more common to identify people by personal attributes (such as job or gender) than by kinship or their hometown.

3. Mass society theorists **Rolf Dahrendorf** and **Peter Berger** argue that bureaucracies have obtained virtually complete control of the individual's life, and the overall bureaucratization of social life has supported the rise of large government.

C. Social Inequality, Powerlessness, and the Individual

1. According to theorists **Karl Marx** and **Jurgen Habermas**, another product of modernization is pronounced social stratification, which results in greater personal feelings of powerlessness because building a stable personal identity is difficult in a highly modernized society that presents the individual with complex, conflicting choices about how to live.

2. According to Habermas, individuals in highly modernized environments are more likely to experiment with new religions, social movements, and lifestyles in search of a fit with their conception of their own "true self."

3. Social theorist **David Reisman** argued that there are three main orientations of personality that can be traced to social structural conditions.

   a. Modernization tends to produce **other-directedness**, which occurs when the behavior of the individual is guided by the observed behavior of others. It is characterized by rigid conformity and openness to the influences of group pressures, changing styles, and shifting interests.

   b. **Inner-directedness**, more common in less modernized societies, occurs when the individual is guided by internal principles and morals rather than the superficialities of the external environment.

   c. **Tradition-directedness** is characterized by a strong conformity to long-standing and time-honored norms, practices, and styles of life.

4. The inner-directed are less likely to sway with the presence or absence of modernization. Because modernization tends to produce other-directedness, anyone who happens to be inner-directed or tradition-directed in a modernized society such as the United States is likely to be seen as a deviant person.

5. Social theorist **Herbert Marcuse** argued that modernized society fails to meet people's basic needs, including the need for a fulfilling identity and a feeling of control over one's own life, resulting in *alienation* of the individual from society. Those groups who have traditionally been denied access to power, such as racial minorities, women, and the working class, particularly experience this alienation.

V. GLOBAL THEORIES OF SOCIAL CHANGE

Globalization refers to the increased interconnectedness and interdependence of different societies around the world. In Europe, this trend has proceeded as far as a common currency for all nations participating in the newly constructed, common economy. As societies become more interconnected, cultural diffusion between them creates common ground; however, cultural differences may become more important as relationships among nations become more intimate.

A. Modernization Theory
  1. Strongly influenced by functionalist theories of social change, **modernization theory** states that global development is a worldwide process affecting nearly all societies that have been touched by social change.
  2. Recent proponents of modernization theory assert that in addition to the U.S. and western European countries, Japan, Taiwan, and South Korea have also led the process of technological globalization and its resultant homogenization.
  3. As a result of their involvement, Japan, with its cultural emphasis on the importance of small friendship groups in the workplace and a traditional work ethic, has profoundly influenced the organization of work in other countries.

B. World Systems Theory
  1. **World systems theory**, formulated by Immanuel Wallerstein, argues that all nations are members of a worldwide system of unequal political and economic relationships that benefit the more developed, technologically advanced countries at the expense of the less developed, less technologically advanced countries.
  2. Wallerstein noted that the world system consists of *core nations* (such as the U.S. and Japan) that produce goods by importing raw materials and cheap labor from *noncore* (or *peripheral*) *nations* (situated in Africa, Latin and South America, and parts of Asia), which suffer exploitation in the global economy.

C. Dependency Theory

Closely allied with Wallerstein's world systems theory is **dependency theory**, which maintains that highly industrialized nations tend to imprison developing nations in dependent relationships, rather than spurring the upward mobility of developing nations with transfers of technology and business acumen.

  1. Dependency theory views industrialized core nations as transferring only those narrow capabilities that it serves them to deliver.
  2. Once these unequal relationships are forged, core nations seek to preserve the status quo because they derive benefits in the form of cheap raw materials and labor from noncore nations.
  3. The developing nations remain dependent upon the core nations for markets and support to keep what industry they have acquired in working order, while at home they experience minimal social development, limited economic growth, and increased income stratification among their own population.
  4. Robert Reich noted that *borrowing dependency* develops when core nations impose terms on loans to noncore nations that put the noncore nations under severe economic strain.

VI. **COLLECTIVE BEHAVIOR AND SOCIAL MOVEMENTS**

Collective behavior occurs when the usual conventions of behavior are suspended and people establish new norms in response to an emerging situation. Types of collective behavior include crowds, panics, disasters, riots, fads, and fashion. **Social movements** are groups that act with some continuity over time to promote or resist change in society.

A. Characteristics of Collective Behavior

1. *Collective behavior always represents the actions of groups of people, not of individuals.* It is rooted in the relationships between people and the norms governing group behavior.
2. *Collective behavior involves new or emergent relationships that arise in unusual or unexpected circumstances.* It arises when uncertainty in the environment creates the need for new forms of social action.
3. Because of its emergent nature, *collective behavior captures the more novel, dynamic, and changing elements of society to a greater degree than other forms of action.*
4. *Collective behavior may mark the beginnings of more organized social behavior.* It often precedes the establishment of formal social organizations.
5. *Collective behavior is patterned behavior, not the irrational behavior of crazed individuals.* It is relatively coordinated among the participants.
6. *During collective behavior, people communicate extensively through* rumors, which develop when people try to define ambiguous situations because they lack adequate information to interpret a problematic event.

B. The Organization of Social Movements
   1. There is typically tension between spontaneity and structure in social movements, which contain routine elements of organization, yet may have to continually develop new strategies and tactics to accomplish change.
   2. **Personal transformation movements** focus on the development of new meaning within individual lives, rather than pursuing social change. An example is the New Age movement, which promotes relaxation and spiritualism as an emotional release from highly stressful, overly rational mainstream life.
   3. **Social/political change movements** aim to change some aspect of society using a variety of tactics, strategies, and organizational forms to achieve their goals.
       a. **Reform movements** seek change through mainstream political or legal means, such as lobbying for legislation that protects the environment.
       b. **Radical movements** seek fundamental change in the structure of society, often by using more dramatic, disruptive tactics, such as sit-ins.
   4. **Reactionary movements** organize to resist change or reinstate an earlier social order. For example, the militia movement in the U.S. resists government authority and seeks to reinstate the perceived lost power of White people.

C. Origins of Social Movements
   1. The formation of a social movement requires a *pre-existing communication network* that facilitates communication among the people who will participate in the new movement. For example, Rosa Parks, who refused to give up her seat on a Montgomery, Alabama bus in 1955, was the secretary of the local NAACP. When she was arrested, the news spread quickly via an already established network of friends, kin, church members, and school organizations.
   2. There must also be *a perceived sense of injustice among the potential participants.* For example, the environmental justice movement developed from grassroots organizations in communities where toxic waste sites and polluting industries posed a perceived threat to residents' health and safety.

3. The *ability of groups to mobilize* is the third factor needed to initiate a social movement. **Mobilization** is the process by which social movements and their leaders secure and coordinate people and resources for the movement.

D. Theories of Social Movements

1. **Resource mobilization theory** is an explanation of how social movements develop that focuses on how movements gain momentum by successfully garnering and mobilizing resources while competing with other movements for money, communication technology, interpersonal contacts, special technical or legal knowledge, and members with organizational and leadership skills.

2. **Political process theory** posits that movements achieve success by exploiting a combination of internal factors, such as the ability of organizations to mobilize resources, and external factors, such as changes occurring in society. This theory stresses that the political system may become structurally vulnerable to social protest during situations such as war, demographic shifts, or economic crisis.

3. Resource mobilization theory and political process theory are social structural explanations that focus on forces external to individuals, while other theories are more cultural in their focus [Table 16.2].

4. **New social movement theory** conceptually links culture, ideology, and identity to explain how members socially construct new identities through their participation in a social movement. This theory explains the demise of the American Indian Movement as a result of the failure to build a pantribal sense of identity *and* the inability to mobilize resources due to government repression.

VII. DIVERSITY. GLOBALIZATION, AND SOCIAL CHANGE

Social movements are a significant source of social change. With the development of *globalization*, movements originating in a particular country will likely have a worldwide influence. The concept of a **transnational social movement** is another way to think about globalization and social movements. In the United States, some of the most significant social movements are those associated with the nation's diverse population, and trend that is likely to continue. There will also be a continuing tension between these social movements and groups resisting social change. Sociologists use the concept of *framing* to explain the process whereby people in social movements develop a shared definition of the situation. **Frames** are specific schemes of interpretation that allow people to identify, perceive, and label events within their lives that can become the basis for collective action.

INTERNET EXERCISES

1. The World Factbook of the CIA at (www.odci.gov/cia/publications/factbook/) provides excellent and reliable information on most countries in the world. Each country listing includes a map and information on the geography, people, government, economics, transportation, communications, military, and transnational issues of that country. Select a particular country that interests you and summarize the major characteristics of the country. Based on this information, would you categorize the country as a core, semiperipheral, or peripheral country according to world systems theory? Explain.

2. The United Nation Children's Program website (www.unicef.org/) produces regular reports on a variety of issues affecting the world's children, such as land mines, child labor, the girl child, and the effects of armed conflict on children. Review the latest State of the World's Children report, and review the Executive Summary. Based on this report, how would you describe the major issues confronting most of the world's children today?

3. The Southern Poverty Law Center (www.splcenter.org/) is devoted to monitoring and abolishing hate groups in the US. Its monthly Intelligence Report and monitoring of hate groups and hate crimes provide a valuable resource for research, awareness, and activism. See the related site: tolerance.org. Select one hate group and describe how it operates and what information is available about its membership and influence.

INFOTRAC EXERCISES

Conduct a keyword search using InfoTrac College Edition to extend the discussion of PUBLIC SOCIOLOGY: *The 9/11 Commission Report on Terrorist Attacks Against the U.S.* (p. 450-1).

1. Keyword: 9/11 Commission Report. This keyword search leads to articles on the findings of the report, criticisms of the report, the role of whistleblowers and so on

2. Keyword: War on terror. This search yields extensive articles on various aspects of the war on terror, which you can review and select those articles that organize around a theme of interest.

3. Keyword: Politics and 9/11. This search leads to articles about political interpretations and responses to 9/11, as well as domestic policies, erosion of civil liberties and so on, as a result of 9/11.

4. Keyword: Disaster response. This search leads to articles about preparedness for 9/11 as a disaster, and the responses to the hurricanes, Katrina and Rita, and what that says about readiness for a disaster similar to 9/11.

PRACTICE TEST

Multiple Choice Questions

1. Which of the following is a common characteristic of collective behavior?
   a. It reflects the consistent, well-established elements of society.
   b. It involves people acting as a group to establish new norms of behavior.
   c. It is completely unpredictable because it emerges in response to new situations.
   d. It usually results from the actions of a group of overly emotional, irrational individuals.

2. Gradual transformations that occur on a broad scale and affect many aspects of society, such as the rise of the computer and emergence of digital culture, are _____.
   a. microchanges
   b. macrochanges
   c. cultural diffusions
   d. demographic shifts

3. Which one of the following statements about social change is (are) **true**?
   a. Social change often creates conflict in society.
   b. Sociologists can accurately predict all of the consequences of social change.
   c. Social change is a process that affects all segments of society equally and similarly.
   d. all of these choices

4. According to _____ theory, social change is the result of inevitable conflict between social classes in capitalist societies.
   a. symbolic interaction.
   b. functionalist
   c. conflict
   d. cyclical

5. Toynbee's assertion that all societies are born, mature, decay, and sometimes die reflects the principles of the _____ theory of social change.
   a. modernization
   b. functionalist
   c. conflict
   d. cyclical

6. Lenski argues that although other elements contribute to social change, _____ are especially important in facilitating social changes such as alterations in religious preferences, the nature of law, and the form of government in modern societies.
   a. immigration patterns
   b. technological advances
   c. changes in family structure
   d. changes in the age composition of the population

7. Revolutions are likely to occur under which of the following condition(s)?
   a. conflict between an oppressive state and various disenfranchised or dissatisfied groups
   b. the opportunity for dissatisfied groups to mobilize en masse
   c. a major economic crisis
   d. all of these choices

8. White youth from the suburbs regularly listen to rap music, a cultural form that began in urban African American communities. This is an example of cultural _____.
   a. lag
   b. conflict
   c. diffusion
   d. inequality

9. Anthropologists argue that the introduction of steel into the Yanomami Indian communities of Brazil and Venezuela has led to _____.
   a. increase in deadly warfare
   b. tribes' inability to continue farming efficiently
   c. emergence of environmental pollution due to automobile emissions
   d. all of these choices

10. According to Herbert Marcuse, _____.
    a. modern society and its attendant technological advances are stable and rational
    b. the technological advances of modern society increase the feeling of having control over one's life
    c. individuals in modern societies experience feelings of alienation and powerlessness
    d. all of these choices

11. Which of the following is not a feature of modernization?
    a. decline in importance of secondary groups to social interaction
    b. decrease in the number of small, close-knit communities
    c. increase in feelings of uncertainty and powerlessness
    d. increase in the mechanization of daily life

12. According to Tonnies, the kind of social organization characterized by a high division of labor, less prominence of personal ties, the lack of a sense of community among the members of society and the absence of a feeling of belonging is typically found in a(n) _____ society.
    a. agrarian
    b. gesellschaft
    c. gemeinschaft
    d. homogeneous

13. Rumors are especially likely to develop when people _____.
    a. are uncomfortable in large group situations
    b. have low levels of education
    c. lack adequate information
    d. do not question authority

14. The Sierra Club has promoted environmental protection by working within the existing political and legal structures. This is an example of what type of social movement?
    a. reactionary
    b. rational
    c. radical
    d. reform

15. According to Sorokin's cyclical model of social change, the contemporary New Age movement represents _____ culture.
    a. utopian
    b. sensate
    c. ideational
    d. technological

16. According to David Reisman, _____ is the personality orientation whereby individuals are relatively impervious to the superficialities around them and are guided by strong internal moral principles.
    a. inner-directedness
    b. other-directedness
    c. value-directedness
    d. tradition-directedness

17. According to Marcuse, _____, or a feeling of individual powerlessness, is more likely to affect those who have traditionally been denied access to power, such as women and racial minorities.
    a. isolation
    b. alienation
    c. deindividuation
    d. identity disruption

18. Which theory asserts that all nations are members of a worldwide system of unequal political and economic relationships that benefit the highly developed, technologically advanced countries at the expense of the less technologically advanced, less developed countries of the world?
    a. Exploitation theory
    b. Modernization theory
    c. World Systems theory
    d. Global Network theory

19. Nations such as the U.S. and England import raw materials and cheap labor from _____ nations, which are commonly located in Africa, Latin and South America, and parts of Asia.
    a. mass
    b. core
    c. noncore
    d. evolutionary

20. Which theory views social movements as starting when structural weaknesses, such as war or economic crisis, present opportunities for collective behavior to occur?
    a. new social movement theory
    b. resource mobilization theory
    c. political process theory
    d. competition theory

True-False Questions

1. Tonnies observed that both patriarchy and the role of the family are considerably less prominent in a *gesellschaft* than in a *gemeinshaft*.

   TRUE or FALSE

2. Demographers predict that by 2050, Asians will comprise 25 percent of the American population.

   TRUE or FALSE

3. According to dependency theory, industrialized nations loan developing nations money at unusually low interest rates to promote economic and social development in the developing nations.

   TRUE or FALSE

4. According to cyclical theory, the first phase of societal development is characterized by tension between the ideal and the practical.

   TRUE or FALSE

5. Tradition-directedness, or strong conformity to long-standing norms, practices, and lifestyles, is the most common personality orientation found in horticultural and agricultural societies.

   TRUE or FALSE

6. Skocpol's research on Russia provides strong support for Marx's argument that class conflict is the single most important variable in explaining why some nations collapse.

   TRUE or FALSE

7. The level of interpersonal trust in large cities is considerably less than in small towns, and social interaction tends to be even more confined within ethnic, racial, and social class groups.

   TRUE or FALSE

8. It is more common to identify people by kinship and hometown than by personal attributes, such as gender and occupation, in a gemeinschaft society.

   TRUE or FALSE

9. The spread of bungie jumping as a form of recreation represents a macrochange in society.

   TRUE or FALSE

10. With increasing modernization, there is generally a decrease in the social importance of formal religious institutions.

    TRUE or FALSE

Fill-in-the-Blank Questions

1. According to _____ theory, societies change from simple to complex and from an undifferentiated to a highly differentiated division of labor.

2. Sociologists refer to the specific schemes of interpretation that allow people to perceive, identify, and label events in their lives as _____ .

3. The Right to Life Movement in the U.S., which is organized around the goal of overturning the Supreme Court decision, *Roe v. Wade*, is an example of a _____ social movement.

4. According to Sorokin, the third phase societies pass through is _____ culture, which involves the hedonistic and sensual elements of culture.

5. According to Reisman, the personality orientation guided by rigid conformity and attempts to "keep up with the Joneses" in modern society is _____ .

Essay Questions

1. Identify the main characteristics of social change and give an example of each characteristic as it relates to the cyberspace revolution.

2. Explain why the unidimensional evolutionary theory of social change has been strongly criticized, and discuss how functionalist theorists now view social diversity.

3. Identify a significant change in the population and discuss how it affects social relationships and social institutions in the United States.

4. Use dependency theory to explain why developing nations have not been able to achieve mobility within the global economic system.

5. Identify the elements necessary for a social movement to begin, and give specific examples of these elements as they relate to the civil rights movement in the United States.

Solutions

PRACTICE TEST

Multiple Choice Questions

1. B, 454-5 Collective behavior occurs when the usual norms are suspended and people collectively establish new norms in response to an emerging situation. Collective behavior captures the novel, dynamic, and changing elements of society. Although it may appear irrational, it is patterned and coordinated.

2. B, 440 Gradual transformations that occur on a broad scale and affect many aspects of society, such as the rise of the personal computer, are macrochanges. Microchanges are subtle alternations in the daily interaction between people, such as fads. Cultural diffusion is the process of transmission of cultural elements, such as fashion, from one society or cultural group to another. Demographic shifts represent changes in the population.

3. A, 440-1 Social change often creates conflict in society. Social change is uneven, affecting different parts of society to varying degrees and at different rates. The onset and consequences of social change are often unforeseen. The direction of social change is not random; rather, it occurs relative to a society's history.

4. C, 443 Functionalist theorists view societies as moving from structurally simple and homogeneous to more complex and heterogeneous. They view technology as the primary cause of social change. Conflict theorists view social change as the result of the inevitable economic conflict between social classes in capitalist societies. Cyclical theorists view social change as necessary for growth as the society moves through the phases of its life cycle.

5. D, 443 Toynbee's ideas are associated with the cyclical theory of social change, which views societies as moving through a life cycle.

6. B, 444 Lenski's ideas are associated with multidimensional evolutionary theory. He asserts that technological advances are primarily responsible for other changes in modern societies.

7. D, 444 Revolutions can occur when there is a conflict between an oppressive state and various disenfranchised or dissatisfied groups, the ability to mobilize en mass, and a political or economic crisis.

8. C, 444-5 The popularity of rap music with White suburban youth in the United States is an example of cultural diffusion, in which cultural elements from one society or group are transmitted to another.

9. A, 446 The Yanomami tribes do not use electricity or automobiles, and the introduction of steel tools generally makes farming more efficient. Anthropologists argue that warlike behavior in these groups has increased as a result of contact with outsiders, or at least that warfare has become more deadly with the use of steel weapons.

10. C, 452 According to Herbert Marcuse, modern society and its attendant technological advances are unstable and irrational, and technological advances of modern society reduce the feeling of having control over one's life, leading to feelings of powerlessness and alienation.

11. A, 448-9 Modernization is a process of social and cultural change that is initiated by industrialization and followed by increased social differentiation and a more complex division of labor. The importance of small, traditional communities declines, as does the centrality of the religious institution to society. Feelings of uncertainty and powerlessness increase, as does the importance of secondary groups to social interaction.

12. B, 449-50 According to Tonnies, the kind of social organization characterized by a high division of labor, less prominence of personal ties, the lack of a sense of community among the members of society, and the absence of a feeling of belonging is typically found in a gesellschaft society. In a gemeinschaft society, there is more homogeneity, members feel a greater sense of belonging, and personal ties are more important to social interaction.

13. C, 455 When they lack adequate information, people communicate through rumors in an attempt to define an ambiguous situation.

14. D, 456-7 The Sierra Club has promoted environmental protection by placing pressure on existing social and political institutions, making it a reform movement. Radical movements often rely on more disruptive tactics to transform the social structure.

15. C, 444 According to cyclical theories of social change, the first phase is characterized by idealistic culture, in which there is tension between the ideal and practical aspects of society. In the second phase, ideational culture emerges, which emphasizes new forms of spirituality. The third phase, sensate culture, stresses practical approaches to reality and involves the hedonistic and the sensual.

16. A, 451 According to Reisman, other-directedness occurs when the individual's behavior is guided by the observed behavior of others. Inner-directedness occurs when the individual is guided by strong internal principles. Tradition-directedness is characterized by strong conformity to long-standing norms and practices.

17. B, 452 Marcuse argues that alienation, or a feeling of powerlessness, is more likely to affect women, racial-ethnic minorities, and the working class, groups traditionally denied access to power in society.

18. C, 453 World systems theory argues that all nations are members of a worldwide system of unequal political and economic relationships that benefit the developed and technologically advanced countries at the expense of the less technologically advanced and less developed countries. Modernization theory states that global development is a worldwide process that affects nearly all societies. Functionalist theory views societies as evolving from structurally simple, homogeneous societies to more complex, highly differentiated societies.

19. C, 453 World systems theory asserts that core nations such as the U.S. import raw materials and cheap labor from non-core nations located in less developed areas of the world. Dependency theory asserts that this arrangement exploits non-core nations by imposing dependency to benefit core nations economically.

20. C, 459 Political process theory argues that movements exploit structural opportunities, such as war, which make political systems vulnerable to social protest.

True-False Questions

1. T, 449 Tonnies observed that both patriarchy and the role of the family are less prominent in a gesellschaft than in a gemeinshaft.

2. F, 447 Demographers predict that Hispanics will be 25 percent and Asians will comprise 9 percent of the U.S. population by 2050.

3. F, 453 Dependency theory asserts that industrialized nations keep developing nations dependent through trade and borrowing practices, often on *high* interest rates that put severe economic strain on these nations.

4. T, 444 The society wrestles with the tension between the ideal and the practical in the idealistic culture, which represents the first phase of societal development according to cyclical theory.

5. T, 452 Tradition-directedness is common in less modernized gemeinschafts.

6. F, 443 Skocpol's research indicates that the relationship *between* societies is a critical factor in explaining why some nations collapse.

7. T, 450 The level of interpersonal trust is considerably less in large cities than in small towns, and social interaction tends to be even more confined within ethnic, racial, and social class groups.

8. T, 450 In gesellschaft societies, people are more likely to be identified by occupation, while in gemeinschaft societies, people are more likely to be identified by kinship group.

9. F, 440 The spread of bungie jumping represents a subtle alteration in the daily interaction between people, rather than a broad social transformation. It is thus an example of a microchange.

10. T, 450-1 Modernization results in the decline in the importance of the religious institution and the prevalence of small, close-knit communities. The role of the government in people's lives increases, as does the importance of the mass media.

Fill-in-the-Blank Questions

1. functionalist, 444
2. frames, 459
3. reactionary, 457
4. sensate, 446
5. other-directed, 452

Essay Questions

1. Cyberspace Revolution and Social Change. The cyberspace revolution is discussed on p. 446 and the main characteristics of social change are described on p. 440-441 in the text. Student responses should demonstrate an understanding of the characteristics of the cyberspace revolution and how it reflects the main characteristics of social change in general.

2. Evolutionary Theory. The unidimensional and multidimensional theories of social change are discussed in the text on p. 441-442. Student responses should demonstrate understanding of how functionalist theories emphasize the role of racial-ethnic, social class, and gender differences in the process of social change.

3. Population and Change. The impact of population or demographic trends on social change is discussed in the text on p. 447. Student responses should demonstrate ability to use a specific example of population change and show how it produced social changes.

4. Dependency Theory. Depending theory is discussed in the text on p. 453. Student responses should demonstrate an understanding of how dependency theory explains the position of development nations within the global economic system and why they have not been able to achieve mobility.

5. Social Movements. The elements necessary for a social movement to being are discussed in the text on p. 457-458. Student responses should demonstrate an understanding of these elements and an ability to apply them to the emergency of the civil rights movement in the United States.

Notes

Notes

Notes

Notes

Notes

Notes

Notes

Notes

Notes

Notes